Connected

Connected

The Surprising Power of
Our Social Networks and
How They Shape Our Lives

Nicholas A. Christakis, MD, PhD

James H. Fowler, PhD

LITTLE, BROWN AND COMPANY
NEW YORK   BOSTON   LONDON

Little, Brown and Company
Hachette Book Group
237 Park Avenue, New York, NY 10017
www.hachettebookgroup.com

First Edition: September 2009

Little, Brown and Company is a division of Hachette Book Group, Inc. The Little, Brown name and logo are trademarks of Hachette Book Group, Inc.

Library of Congress Cataloging-in-Publication Data

Christakis, Nicholas A.
    Connected : the surprising power of our social networks and how they shape our lives / Nicholas A. Christakis and James H. Fowler. — 1st ed.
        p. cm.
    Includes bibliographical references and index.
    ISBN 978-0-316-03614-6 (hc) / 978-0-316-07258-8 (int'l ed)
    1. Social networks.   I. Fowler, James H.   II. Title.
    HM741.C46 2009
    302.3 — dc22   2009018349

10 9 8 7 6 5 4 3 2 1

RRD-IN

Design by Meryl Sussman Levavi
Printed in the United States of America

For Erika, Sebastian, Lysander, and Eleni

and

for Harla, Lucas, and Jay

to whom our connection is aeonian

Contents

Preface

Social networks are intricate things of beauty. They are so elaborate and so complex—and so ubiquitous, in fact—that one has to wonder what purpose they serve. Why are we embedded in them? How do they form? How do they work? How do they affect us? How do they affect us?

I (Nicholas) have been animated by these questions for the better part of the past ten years. I began by being interested in the simplest social network of all: a pair of people, a dyad. Initially, the dyads I studied were husbands and wives. As a physician caring for terminally ill patients and their families, I noticed the serious toll that a loved one's death had on a spouse. And I became interested in how illness in one person might cause illness in another. For it seemed to me that if people are interconnected, their health must also be interconnected. If a wife falls ill or dies, her husband's risk of death assuredly rises. Eventually, I began to realize that there were all kinds of dyads I might study, such as pairs of siblings or pairs of friends or pairs of neighbors who are connected (*not* separated) by a backyard fence.

But the intellectual heart of the matter was not in these simple arrangements. Instead, the key realization was that these dyads agglomerate to form huge webs of ties stretching far into the distance. A man's wife has a best friend who has a husband who has a coworker who has a sibling who has a friend, and so on and so forth. These chains branch like lightning bolts, forming intricate patterns throughout human society. The situation, it seemed, was much more complicated. With every step away from an individual in a social network that we take, the number of ties to other humans, and the complexity of the branching, rise very, very fast. As I reflected on this problem, I began to read the work of other social scientists, from lonely German scholars at the turn of the twentieth century to visionary sociologists in the 1970s, who had studied social networks ranging in size from three to thirty people. But my interest lay in social networks of three thousand or thirty thousand or even three million people.

I realized that to study things of such complexity, I would make better progress if I worked with another investigator. As it turned out, James Fowler, also at Harvard, was studying networks from a completely different perspective. James and I did not know each other despite working in adjoining buildings on the same campus for several years. In 2002 we were introduced by a mutual colleague, political scientist Gary King. In other words, we started our journey as friends of a friend. Gary thought we might have common intellectual interests, and he was right. Indeed, the very fact that we met because of our social network illustrates a major point we want to make about how and why social networks operate and how they benefit us.

James had spent a number of years studying the origin of people's political beliefs and examining how one person's attempt to solve a social or political problem influenced others. How did humans come together to accomplish what they could not do on their own? And he shared interests in other topics that were a key part of the story:

altruism and goodness, both of which are essential for social networks to grow and endure.

Together, as we began to think about the idea that people are connected in vast social networks, we realized that social influence does not end with the people we know. If we affect our friends, and they affect their friends, then our actions can potentially affect people we have never met. We began by studying various health effects. We discovered that if your friend's friend's friend gained weight, you gained weight. We discovered that if your friend's friend's friend stopped smoking, you stopped smoking. And we discovered that if your friend's friend's friend became happy, you became happy.

Eventually, we realized that there were fundamental rules that governed both the formation and the operation of social networks. We concluded that if we were going to study how networks function, we also had to understand how they are assembled. One cannot, for example, be friends with absolutely anybody. People are constrained by geography, socioeconomic status, technology, and even genes to have certain kinds of social relationships and to have a certain number of them. The key to understanding people is understanding the ties between them; therefore, it was to the ties that we turned our focus.

Our interest in these topics paralleled the interests of many other scholars who have advanced the mathematics and science of networks over the past ten years. As we began to study human connections, we encountered engineers studying networks of power stations, neuroscientists studying networks of neurons, geneticists studying networks of genes, and physicists studying networks of darn near everything. Their networks might be pretty too, we thought, but ours were more interesting: much more complicated and much more consequential. After all, the nodes in our networks are thinking human beings. They can make decisions, potentially changing their networks even while embedded in them and being affected by them. A network of humans has a special kind of life of its own.

Just as scientists have become interested in the underlying beauty and explanatory power of networks, the person on the street thinks about them too. This is largely due to the appearance of the Internet in people's homes, which has given everyone a notion of how lots of things might be interconnected. People began to speak colloquially about the "Net" and eventually the "World Wide Web" (not to mention the smash-hit movie *The Matrix*). And they began to realize that they were as interconnected as their computers. These connections have become explicitly social to the point that today nearly everyone is familiar with social-network websites like Facebook and MySpace.

As we studied social networks more deeply, we began to think of them as a kind of human superorganism. They grow and evolve. All sorts of things flow and move within them. This superorganism has its own structure and a function, and we became obsessed with understanding both.

Seeing ourselves as part of a superorganism allows us to understand our actions, choices, and experiences in a new light. If we are affected by our embeddedness in social networks and influenced by others who are closely or distantly tied to us, we necessarily lose some power over our own decisions. Such a loss of control can provoke especially strong reactions when people discover that their neighbors or even strangers can influence behaviors and outcomes that have moral overtones and social repercussions. But the flip side of this realization is that people can transcend themselves and their own limitations. In this book, we argue that our interconnection is not only a natural and necessary part of our lives but also a force for good. Just as brains can do things that no single neuron can do, so can social networks do things that no single person can do.

For decades, even centuries, serious human concerns, such as whether a person will live or die, be rich or poor, or act justly or unjustly, have been reduced to a debate about individual versus collective responsibility. Scientists, philosophers, and others who study society have generally divided into two camps: those who think individuals are in control of their destinies, and those who believe that social forces (ranging from a lack of good public education to the presence of a corrupt government) are responsible for what happens to us.

However, we think that a third factor is missing from this debate. Given our research and our own diverse experiences in life—from meeting our spouses to meeting each other, from caring for terminally ill patients to building latrines in poor villages—we believe that our connections to other people matter most, and that by linking the study of individuals to the study of groups, the science of social networks can explain a lot about human experience. This book focuses on our ties to others and how they affect emotions, sex, health, politics, money, evolution, and technology. But most of all it is about what makes us uniquely human. To know who we are, we must understand how we are connected.

Connected

CHAPTER I

In the Thick of It

In the mountain village of Levie, Corsica, during the 1840s, Anton-Claudio Peretti became convinced that his wife, Maria-Angelina, was having an affair with another man and that, even worse, their daughter was not his child. Maria told Anton that she was going to leave him, and she made preparations to do so with her brother, Corto. That very evening, Anton shot his wife and daughter to death and fled to the mountains. The bereft Corto sorely wanted to kill Anton, but he could not find him. In a bit of violent symmetry that seemed sensible to residents of the area, Corto instead killed Anton's brother, Francesco, and nephew, Aristotelo.

It did not end there. Five years later, Giacomo, brother of the deceased Aristotelo, avenged the deaths of his brother and father by killing Corto's brother. Giacomo wanted to kill Corto's father too, but he had already died of natural causes, denying Giacomo the satisfaction.[1] In this cascade of death, Giacomo and Corto's brother were connected by quite a path: Giacomo was the son of Francesco, who was the brother of Anton, who was married to Maria, who was the sister of Corto, whose brother was the target of Giacomo's murderous wrath.

Such behavior is not restricted to historically or geographically distant places. Here is another example, closer to home: Not long before the summer of 2002 in St. Louis, Missouri, Kimmy, an exotic dancer, left a purse containing $900 in earnings with a friend while she was busy. When she came back to reclaim it, her friend and the purse were gone. But a week later, Kimmy's cousin spotted the purse thief's partner at a local shop, and she called Kimmy. Kimmy raced over with a metal pole. She viciously attacked this friend of her erstwhile friend. Later she observed with pride that she had "beat her [friend's] partner's ass....I know I did something...[to get even] that's the closest thing I could [do]."[2]

Cases like these are puzzling. After all, what did Anton's brother and nephew and Kimmy's friend's friend have to do with anything? What possible sense is there in injuring or killing the innocent? Even by the incomprehensible standards of murderous violence, what is the point of these actions, taken one week or five years later? What explains them?

We tend to think of such cases as quaint curiosities, like Appalachian feuds, or as backward practices, like the internecine violence between Shiite and Sunni tribesmen or the cycle of killings in Northern Ireland or the reciprocating gang violence in American cities. But this grim logic has ancient roots. It is not just that the impetus to revenge is ancient, nor even that such violence can express group solidarity ("we are Hatfields, and we hate McCoys"), but that violence—in both its minor and extreme forms—can spread through social ties and has done so since humans emerged from the African savanna. It can spread either in a directed fashion (retaliating against the perpetrators) or in a generalized fashion (harming nondisputants nearby). Either way, however, a single murder can set off a cascade of killings. Acts of aggression typically diffuse outward from a starting point—like a bar fight that begins when one man swings at another who ducks, resulting in a third man getting hit, and soon (in what has become a cliché precisely because it evokes deep-seated notions of unleashed aggression) punches are flying everywhere. Sometimes these epidemics of violence, whether in Mediterranean villages or urban gangs, can persist for decades.[3]

Notions of collective guilt and collective revenge that underlie cascades of violence seem strange only when we regard responsibility as a personal attribute. Yet in many settings, morality resides in groups rather than in individuals. And a further clue to the collective nature of violence is that it tends to be a public, not a private, phenomenon. Two-thirds of the acts of interpersonal violence in the United States are witnessed by third parties, and this fraction approaches three-fourths among young people.[4]

Given these observations, perhaps the person-to-person spread of violence should not surprise us. Just as it is often said that "the friend of my friend is my friend" and "the enemy of my enemy is my friend," so too the friend of my enemy is my enemy. These aphorisms encapsulate certain truths about animosity and affection, but they also convey a fundamental aspect of our humanity: our connection. While Giacomo and Kimmy acted alone, their actions show just how easily responsibility and retaliation can diffuse from person to person to person across social ties.

In fact, we do not even have to search for complicated paths across which violence spreads, because the initial step, from the very first person to the next, accounts for most of the violence in our society. In trying to explain violence, it is myopic to focus solely on the perpetrator—his frame of mind, his finger on the trigger—because murder is rarely a random act between strangers. In the United States, 75 percent of all homicides involve people who knew each other, often intimately, prior to the murder. If you want to know who might take your life, just look at the people around you.

But your social network also includes those who might save your life. "On March 14, 2002, I gave my right kidney to my best friend's husband," Cathy would later note in an online forum that chronicles the experiences of people who become "living donors" of organs. The summer before, during a heartfelt chat, Cathy had learned that her friend's husband's renal failure had worsened and that he needed a kidney transplant in order to survive. Overcome with the desire to help, Cathy underwent a series of medical and psychological evaluations, getting more and more excited as she passed each one and moved closer to her goal of donating one of her kidneys. "The experience has been the most rewarding of my life," she wrote. "I am so grateful that I was able to help my best friend's husband. His wife has her husband back. His sons have their dad back.... It's a win-win situation. We all win. I gave the gift of life."[5]

Similar stories abound, and such "directed donations" of organs can even come to involve people who have rather tenuous connections, a Starbucks clerk and his longtime customer, for example. There can even be organ-donation cascades that loosely resemble the Perettis' murder cascade. John Lavis, a sixty-two-year-old resident of the town of Mississauga, Ontario, father of four and grandfather of three, was dying of heart failure in 1995. His heart had failed during triple-bypass surgery, and he was placed on a temporary artificial heart. In a stroke of unbelievable good fortune, a donor heart was transplanted into him just eight days later when he was on the brink of death. His daughter recalled: "We were a family of immense gratitude.... [My father] received the biggest gift he will ever receive—his life was given back to him." Motivated by this experience, Lavis's children all signed organ-donor cards, thinking that this symmetrical act was the least they could do. Then in 2007, Lavis's son Dan died in a work-related accident. Eight people benefited from Dan's decision to donate his organs. The woman who received his heart later wrote to the Lavis family, thanking them for "giving her a new life."[6] The same year in the United States, a similar cascade an amazing ten links long took place between unrelated living kidney donors (albeit with explicit medical coordination), saving many lives along the way.[7]

Social-network ties can—and, as we will see, usually do—convey benefits that are the very opposite of violence. They can be conduits for altruistic acts in which individuals pay back a debt of gratitude by paying it forward. The role that social connections can play in the spread of both good and bad deeds has even prompted the creation of novel strategies to address social problems. For example, programs in several U.S. metropolitan areas involve teams of "violence interrupters." These streetwise individuals, often former gang members, try to stop the killing by attempting to break the cycle of transmission. They rush to the bedsides of victims or to the homes of victims' families and friends, encouraging them not to seek revenge. If they can persuade just one person not to be violent, quite a few lives can be saved.

Our connections affect every aspect of our daily lives. Rare events such as murder and organ donation are just the tip of the iceberg. How we feel, what we know, whom we marry, whether we fall ill, how much money we make, and whether we vote all depend on the ties that bind us. Social networks spread happiness, generosity, and love. They are always there, exerting both subtle and dramatic influence over our choices, actions, thoughts, feelings, even our desires. And our connections do not end with the people we know. Beyond our own social horizons, friends of friends of friends can start chain reactions that eventually reach us, like waves from distant lands that wash up on our shores.

Bucket Brigades and Telephone Trees

Imagine your house is on fire. Luckily, a cool river runs nearby. But you are all alone. You run back and forth to the river, bucket in hand, toting gallon after gallon of water to splash on your burning home. Unfortunately, your efforts are useless. Without some help, you will not be able to carry water fast enough to outpace the inferno.

Now suppose that you are not alone. You have one hundred neighbors, and, lucky for you, they all feel motivated to help. And each one just happens to have a bucket. If your neighbors are sufficiently strong, they can run back and forth to the river, haphazardly dumping buckets of water on the fire. A hundred people tossing water on your burning house is clearly better than you doing it by yourself. The problem is that once they get started your neighbors waste a lot of time running back and forth. Some of them tire easily; others are uncoordinated and spill a lot of water; one guy gets lost on his way back to your house. If each person acts independently, then your house will surely be destroyed.

Fortunately, this does not happen because a peculiar form of social organization is deployed: the bucket brigade. Your hundred neighbors form a line from the river to your house, passing full buckets of water toward your house and empty buckets toward the river. Not only does the bucket brigade arrangement mean that people do not have to spend time and energy walking back and forth to the river; it also means that weaker people who might not be able to walk or carry a heavy bucket long distances now have something to offer. A hundred people taking part in a bucket brigade might do the work of two hundred people running haphazardly.

But why exactly is a group of people arranged this way more effective than the same group of people—or even a larger group— working independently? If the whole is greater than the sum of its parts, how exactly does the whole come to be greater? Where does the "greater" part come from? It's amazing to be able to increase the effectiveness of human beings by as much as an order of magnitude simply by arranging them differently. But what is it about combining people into groups with *particular configurations* that makes them able to do more things and different things than the individuals themselves?

To answer these questions, and before we get to the fun stuff, we first need to explain a few basic terms and ideas of network theory. These basic concepts set the stage for the individual stories and the more complicated ideas we will soon explore as we investigate the surprising power of social networks to affect the full spectrum of human experience.

We should first clarify what we mean by a group of people. A *group* can be defined by an attribute (for example, women, Democrats, lawyers, long-distance runners) or as a specific collection of individuals to whom we can literally point ("those people, right over there, waiting to get into the concert"). A social network is altogether different. While a network, like a group, is a collection of people, it includes something more: a specific set of connections between people in the group. These ties, and the particular pattern of these ties, are often more important than the individual people themselves. They allow groups to do things that a disconnected collection of individuals cannot. The ties explain why the whole is greater than the sum of its parts. And the specific pattern of the ties is crucial to understanding how networks function.

The bucket brigade that saves a house is a very simple social network. It is linear and has no branches: each person (except the first and last) is connected to two other people, the one in front and the one behind. For moving something like water long distances, this is a good way to be organized. But the optimal organization of one hundred people into a network depends very much on the task at hand. The best pattern of connections between a hundred people to put out a fire is different from the best pattern for, say, achieving a military objective. A company of one hundred soldiers is typically organized into ten tightly interconnected squads of ten. This allows each soldier to know all of his squad mates rather than just the grunt in front of him and the grunt behind him. The military goes to great lengths to help squad members know each other very well, so well in fact that they are willing to give their lives for one another.

Consider still another social network: the telephone tree. Suppose you need to contact a hundred people quickly to let them know that school is canceled. Before modern communications and the Internet, this was a challenge because there was no public source of up-to-the-minute information that everyone could access from their homes (though the ringing of church bells in the town square comes to mind). Instead, each person needed to be contacted directly. The telephone made this task much easier, but it was still a burden for one person to make all one hundred calls. And even if someone set out to do this, it might take quite a while to get to the people at the end of the list, by which time they may have already left home for school. Having a single person make all the calls is both inefficient and burdensome.

Ideally, one person would set off a chain reaction so that everyone could be reached as quickly as possible and with the least burden on any particular individual. One option is to create a list and have the person at the top of the list call the next person, the second person call the third, and so on until everyone gets the message, as in a bucket brigade. This would distribute the burden evenly, but it would still take a really long time for the hundredth person to be reached. Moreover, if someone in the sequence was not home when called, everyone later in the list would be left in the dark.

An alternative pattern of connections is a telephone tree. The first person calls two people, who each call two people, and so on until everyone is contacted. Unlike the bucket brigade, the telephone tree is designed to spread information to many people simultaneously, creating a cascade. The workload is distributed evenly among all group members, and the problem caused by one person not being home is limited. Moreover, with a single call, one person can set off a chain of events that could influence hundreds or thousands of other people — just as the person who donated the heart that was transplanted into John Lavis prompted another donation that saved eight more lives. The telephone tree also vastly reduces the number of steps it takes for information to flow among people in the group, minimizing the chance that the message will be degraded. This particular network structure thus helps to both amplify and preserve the message. In fact, within a few decades of the widespread deployment of home-based phones in the United States, telephone trees were used for all sorts of purposes. An article in the *Los Angeles Times* from 1957, for example, describes the use of a phone tree to mobilize amateur astronomers, as part of the "Moonwatch System" of the Smithsonian Astrophysical Observatory, to track American and Russian satellites.[8]

Alas, this same network structure also allows a single swindler to cheat thousands of people. In Ponzi schemes, money flows "up" a structure like a telephone tree. As new people are added to the network, they send money to the people "above" them and then new members are recruited "below" them to provide more money. As time passes, money is collected from more and more people. In what might be the biggest Ponzi scheme of all time, federal investigators discovered in 2008 that during the previous thirty years Bernie Madoff had swindled $50 billion from thousands of investors. Like the Corsican vendetta network we described earlier, Madoff's investment network is the kind most of us would like to avoid.

The four different types of networks we have considered so far are shown in the illustration. First is a group of one hundred people (each represented by a circle, or *node*) among whom there are no ties. Next is a bucket brigade. Here, in addition to the one hundred people, there are a total of ninety-nine ties between the members of the group; every person (except the first and last) is connected to two other people by a *mutual tie* (meaning that full and empty buckets pass in both directions). In the telephone tree, there are one hundred people and again ninety-nine ties. But here, everyone, with the exception of the first and last people in the tree, is connected to three other people, with one inbound tie (the person they get the call from) and two outbound ties (the people they make calls to). There are no mutual ties; the flow of information is directional and so are the ties between people. In a company of one hundred soldiers, each member of each squad knows every other member of the squad very well; and each person has exactly nine ties. Here, there are one hundred people and 450 ties connecting them. (The reason there are not nine hundred ties is that each tie counts once for the two people it connects.) In the drawing, we imagine that there are no ties between

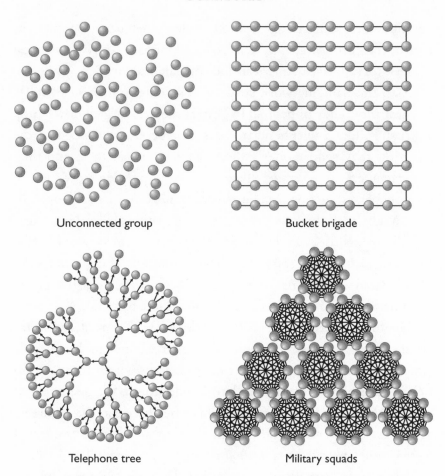

Unconnected group

Bucket brigade

Telephone tree

Military squads

Four different ways to connect one hundred people. Each circle ("node") represents a person, and each line ("tie") a relationship between two people. Lines with arrows indicate a directed relationship; in the telephone tree, one person calls another. Otherwise, ties are mutual: in the bucket brigade, full and empty buckets travel in both directions; in military squads, the connections between the soldiers are all two-way.

squads or, at least, that the ties within squads are much tighter than the ties between squads. This is clearly an oversimplification, but it illustrates still another point about communities in social networks. A *network community* can be defined as a group of people who are much more connected to one another than they are to other groups of connected people found in other parts of the network. The communities are defined by structural connections, not necessarily by any particular shared traits.

In a very basic sense, then, a social network is an organized set of people that consists of two kinds of elements: human beings and the connections between them. Unlike the bucket brigade, telephone tree, and military company, however, the organization of natural social networks is typically not imposed from the top. Real, everyday social networks evolve organically from the natural tendency of each person to seek out and make many or few friends, to have large or small families, to work in personable or anonymous workplaces.

For example, in the next illustration, we show a network of 105 students in a single dormitory at an American university and the friendship ties between them. On average, each student is connected to six other close friends, but some students have only one friend, and others have many. Moreover, some students are more embedded than others, meaning they have more connections to other people in the network via friends or friends of friends. In fact, network visualization software is designed to place those who are more interconnected in the center and those who are less interconnected at the periphery, helping us to see each person's location in the network. When your friends and family become better connected, it increases your level of connection to the whole social network. We say it makes you more *central* because having better-connected friends literally moves you away from the edges and toward the center of a social network. And we can measure your centrality by counting not just the number of your friends and other contacts but also by counting your friends' friends, and their friends, and so on. Unlike the bucket brigade where everyone feels his position to be the same ("there's a guy on my left passing me buckets and a guy on my right to whom I give them—it doesn't matter where in the line I am"), here, people are located in distinctly different kinds of places within the network.

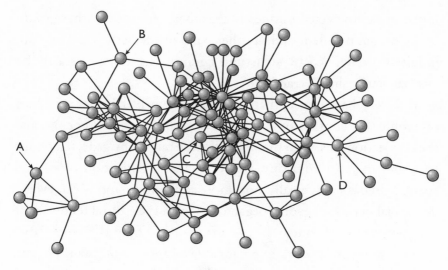

In this natural network of close friendships among 105 college students living in the same dormitory, each circle represents a student, and each line a mutual friendship. Even though A and B both have four friends, A's four friends are more likely to know one another (there are ties between them), whereas none of B's friends know each other. A has greater transitivity than B. Also, even though C and D both have six friends, they have very different locations in the social network. C is much more central, and D is more peripheral; C's friends have many friends themselves, whereas D's friends tend to have few or no friends.

A network's *shape*, also known as its structure or topology, is a basic property of the network. While the shape can be visualized, or represented, in different ways, the actual pattern of connections that determines the shape remains the same regardless of how the network is visualized. Imagine a set of five hundred buttons strewn on the floor. And imagine that there are two thousand strings we can use to connect the buttons. Next, imagine that we randomly select two buttons and connect them with a string, knotting each button at the end. Then we repeat this procedure, connecting random pairs of buttons one after another, until all the strings are used up. In the end, some buttons will have many strings attached to them, and others, by chance, will never have been picked and so will not be connected to another button. Perhaps some groups of buttons will be connected to each other but separated from other groups. These groups—even those that consist of a single unconnected button—are called *components* of the network; when we illustrate networks, we frequently represent only the largest component (in this case, the one with the most buttons).

If we were to select one button from one component and pick it up off the floor, all other buttons attached to it, directly or indirectly, would also be lifted into the air. And if we were to drop this mass of buttons onto another spot on the floor, it would look different than it did when we first picked it up. But the topology—which is a fundamental and intrinsic property of the network of buttons—would be exactly the same, no matter how many times we picked up and dropped the mass of connected buttons. Each button has the same relational position to other particular buttons that it had before; its *location* in the network has not changed. Visualization software tries to show this in two dimensions and to reveal the underlying topology by putting the most tangled buttons in the center and the least connected ones on the edges. It's as if you were trying to untangle a gnarled set of Christmas-tree lights, and there were tendrils of the gnarled mess that you could pull out, and also a thicket of inter-knotted lights that remained in the center no matter how often you turned the tangle over on the floor.

For numerous reasons we will explore, people come to occupy particular spots in the naturally occurring and continuously evolving social networks that surround us. Organic networks have a structure, complexity, function, spontaneity, and sheer beauty not found in organized networks, and their existence provokes questions about how they arise, what rules they obey, and what purpose they serve.

Rules of Life in the Network

There are two fundamental aspects of social networks, whether they are as simple as a bucket brigade or as complex as a large multigenerational family, a college dormitory, an entire community, or the worldwide network that links us all. First, there is *connection*, which has to do with who is connected to whom. When a group is constituted as a network, there is a particular pattern of ties that connects the people involved, the topology. Moreover, ties are complicated. They can be ephemeral or lifelong; they can be casual or intense; they can be personal or anonymous. How we construct or visualize a network depends on how we define the ties of interest. Most analyses emphasize ties to family, friends, coworkers, and neighbors. But there are all sorts of social ties and, thus, all sorts of social networks. In fact, when things such as sexually transmitted diseases or dollar bills flow through a network, this flow itself can define the ties and hence the structure of a particular set of network connections.

Second, there is *contagion*, which pertains to what, if anything, flows across the ties. It could be buckets of water, of course, but it also could be germs, money, violence, fashions, kidneys, happiness, or obesity. Each of these flows might behave according to its own rules. For example, fire cannot be transported in buckets toward the river; germs cannot affect someone who is immune; and obesity, which we will discuss in chapter 4, tends to spread faster between people of the same sex.

Understanding why social networks exist and how they work requires that we understand certain rules regarding connection and contagion—the structure and function—of social networks. These principles explain how ties can cause the whole to be greater than the sum of the parts.

RULE 1: WE SHAPE OUR NETWORK

Humans deliberately make and remake their social networks all the time. The primary example of this is *homophily,* the conscious or unconscious tendency to associate with people who resemble us (the word literally means "love of being alike"). Whether it's Hells Angels or Jehovah's Witnesses, drug addicts or coffee drinkers, Democrats or Republicans, stamp collectors or bungee jumpers, the truth is that we seek out those people who share our interests, histories, and dreams. Birds of a feather flock together.

But we also choose the *structure* of our networks in three important ways. First, we decide how many people we are connected to. Do you want one partner for a game of checkers or many partners for a game of hide-and-seek? Do you want to stay in touch with your crazy uncle? Do you want to get married, or would you rather play the field? Second, we influence how densely interconnected our friends and family are. Should you seat the groom's college roommate next to your bridesmaid at the wedding? Should you throw a party so all your friends can meet each other? Should you introduce your business partners? And third, we control how central we are to the social network. Are you the life of the party, mingling with everyone at the center of the room, or do you stay on the sidelines?

Diversity in these choices yields an astonishing variety of structures for the whole network in which we come to be embedded. And it is diversity in these choices—a diversity that has both social and genetic origins as we will see in chapter 7—that places each of us in a unique location in our own social network. Of course, sometimes these structural features are not a matter of choice; we may live in places that are more or less conducive to friendship, or we may be born into large or small families. But even when these social-network structures are thrust upon us, they still rule our lives.

We actually know quite a bit about how people vary in terms of how many friends and social contacts they have and in how interconnected they are. Yet, identifying who a person's social contacts are can be a tricky business since people have many interactions of varying intensities with all sorts of people. While a person may know a few hundred people by sight and name, he will typically be truly close to only a few. One way social scientists identify such close individuals is to ask questions like, who do you discuss important matters with? Or, who do you spend your free time with? When answering such questions, people will identify a heterogeneous mix of friends, relatives, coworkers, schoolmates, neighbors, and others.

We recently put these questions to a sample of more than three thousand randomly chosen Americans. And we found that the average American has just four close social contacts, with most having between two and six. Sadly, 12 percent of Americans listed no one with whom they could discuss important matters or spend free time. At the other extreme, 5 percent of Americans had eight such people. About half of the people listed as members of Americans' intimate groups were said to be friends, but the other half included a wide variety of different kinds of relationships, including spouses, partners, parents, siblings, children, coworkers, fellow members of clubs, neighbors, and professional advisers and consultants. Sociologist Peter Marsden has called this group of people that we all have a "core discussion network." In a national sample of 1,531 Americans studied in the 1980s, he found that core-discussion-network size decreases as we age, that there is no overall difference between men and women in core-network size, and that those with a college degree have core networks that are nearly twice as large as those who did not finish high school.[9]

Next, in our own work, we asked the respondents to tell us how interconnected their social contacts were to each other. So if a person said that Tom, Dick, Harry, and Sue were his friends, we asked him if Tom knew Dick, if Tom knew Harry, if Tom knew Sue, if Dick knew Harry, and so on. We then used these answers to calculate the probability that any two of a person's friends were also friends with each other. This probability is an important property that we use to measure how tightly interwoven a network is.

If you know Alexi, and Alexi knows Lucas, and Lucas knows you, we say this relationship is *transitive*—the three people involved form a triangle. Some people live in the thick of many transitive relationships (like person A in the illustration on page 14), while others have friends who do not know each other (like person B). Those with high transitivity are usually deeply embedded within a single group, while those with low transitivity tend to make contact with people from several different groups who do not know one another, making them more likely to act as a bridge between different groups. Overall, we found that if you are a typical American, the probability that any two of your social contacts know each other is about 52 percent.

Although these measures characterize the networks we can see, they also tell us something about the networks we cannot see. In the vast fabric of humanity, each person is connected to his friends, family, coworkers, and neighbors, but these people are in turn connected to their friends, family, coworkers, and neighbors, and so on endlessly into the distance, until everyone on earth is connected (pretty much) to everyone else, one way or another. So whereas we think of our own network as having a more limited social and geographic reach, the networks that surround each of us are actually very widely interconnected.

It is this structural feature of networks that underlies the common expression "it's a small world." It is often possible, through a few connections from person to person, for an individual to discover a connection to someone else. A famous example (at least among social scientists) was described in a paper first drafted in the 1950s by two early figures in the study of social networks, Ithiel de Sola Pool and Manfred Kochen. One of the authors overheard a patient in a hospital in a small town in Illinois say to a Chinese patient in the adjoining bed: "You know, I've only known one Chinese before in my life. He was——from Shanghai." Whereupon the response came back, "Why, that's my uncle."[10] In fact, the authors did not tell us his name, perhaps because they were worried that the reader, in a further illustration of the small-world effect, would know him.

RULE 2: OUR NETWORK SHAPES US

Our place in the network affects us in turn. A person who has no friends has a very different life than one who has many. For example, we will see in chapter 4 that having an extra friend may create all kinds of benefits for your health, even if this other person doesn't actually do anything in particular for you.

One study of hundreds of thousands of Norwegian military conscripts provides a simple example of how the mere number of social contacts (here, siblings) can affect you.[11] It has been known for some time that first-born children score a few points higher in terms of intelligence than second-born children, who in turn score a bit higher than third-born children. One of the outstanding questions in this area of investigation, however, has been whether these differences are due to biological factors fixed at birth or to social factors that come later. The study of Norwegian soldiers showed that simple features of social networks, such as family size and structure, are responsible for the differences. If you are a second-born son whose older sibling died while you were a child, your IQ increases and resembles the IQ of a first-born child. If you are a third-born child and one of your older siblings died, your IQ resembles that of a second-born child; and if both of your older siblings died, then your IQ resembles that of a first-born child.

Whether your friends and other social contacts are friends with one another is also crucial to your experience of life. Transitivity can affect everything from whether you find a sexual partner to whether you commit suicide. The effect of transitivity is easily appreciated by the example of how divorce affects a child. If a child's parents are married (connected) then they probably talk to each other, but if they get divorced (disconnected) they probably do not. Divorce means that communication often has to pass through the child ("Tell your father not to bother picking you up next Saturday!"), and it is much harder to coordinate raising the child ("You mean your mother bought you ice cream too?"). What is remarkable is that even though the child is still deeply connected to both parents, her relationship with each of them changes as a consequence of the divorce. Yet these changes result from the loss of a connection between the parents—a connection the child has little to do with. The child still has two parents, but her life is different depending on whether or not they are connected.

And how many contacts your friends and family have is also relevant. When the people you are connected to become better connected, it reduces the number of hops you have to take from person to person to reach everyone else in the network. You become more central. Being more central makes you more susceptible to whatever is flowing within the network. For example, person C in the figure on page 14 is more central than person D. Ask yourself which person you would rather be if a hot piece of gossip were spreading; you should be person C. Now ask yourself which person you would rather be if a deadly germ were spreading in the network; you should be person D. And this is the case even though persons C and D each have the same number of social ties: they are each directly connected to just six people. In later chapters, we will show how your centrality affects everything from how much money you make to whether you will be happy.

RULE 3: OUR FRIENDS AFFECT US
The mere shape of the network around us is not all that matters, of course. What actually flows across the connections is also crucial. A bucket brigade is formed not to make a pretty line for you to look at while your house is burning but so that people can pass water to each other to douse the flames. And social networks are not

just for water—they transport all kinds of things from one person to another.

As we will discuss in chapter 2, one fundamental determinant of flow is the tendency of human beings to influence and copy one another. People typically have many direct ties to a wide variety of people, including parents and children, brothers and sisters, spouses (and nice ex-spouses), bosses and coworkers, and neighbors and friends. And each and every one of these ties offers opportunities to influence and be influenced. Students with studious roommates become more studious. Diners sitting next to heavy eaters eat more food. Homeowners with neighbors who garden wind up with manicured lawns. And this simple tendency for one person to influence another has tremendous consequences when we look beyond our immediate connections.

RULE 4: OUR FRIENDS' FRIENDS' FRIENDS AFFECT US
It turns out that people do not copy only their friends. They also copy their friends' friends, and their friends' friends' friends. In the children's game telephone, a message is passed along a line by each child whispering into the next child's ear. The message each child receives contains all the errors introduced by the child sharing it as well as those introduced by prior children to whom the child is not directly connected. In this way, children can come to copy others to whom they are not directly tied. Similarly, every parent warns children not to put money in their mouths: the money, we think, contains germs from numerous people whose hands it has passed through, and not just from the most recent pair of hands. Analogously, our friends and family can influence us to do things, like gain weight or show up at the polls. But their friends and family can influence us too. This is an illustration of *hyperdyadic spread,* or the tendency of effects to spread from person to person to person, beyond an individual's direct social ties. Corto's brother lost his life because of such spread.

It is easy to think about hyperdyadic effects when the network is a straight line—("that guy three people down the line better pass the bucket, or we're all going to be in big trouble"). But how on earth can they be understood in a natural social network such as the college students in the illustration on page 14, or complex networks of thousands of people with all kinds of crosscutting paths stretching far beyond the social horizon (as we will consider later)? To decipher what is going on, we need two kinds of information. First, we must look beyond simple, sequential dyads: we need to know about individuals and their friends, their friends' friends, their friends' friends' friends, and so on. And we can only get this information by observing the whole network at once. It has just recently become possible to do this on a large scale. Second, if we want to observe how things flow from person to person to person, then we need information about the ties and the people they connect at more than one point in time, otherwise we have no hope of understanding the dynamic properties of the network. It would be like trying to learn the rules of an unfamiliar sport by looking at a single snapshot of a game.

We will consider many examples and varieties of hyperdyadic spread, but we can set the stage with a simple one. The usual way we think about contagion is that if one person has something and comes into contact with another person, that contact is enough for the second person to get it. You can become infected with a germ (the most straightforward example) or with a piece of gossip or information (a less obvious example). Once you get infected by a single person, additional contact with others is generally redundant. For example, if you have been told accurately that stock XYZ closed at $50, another person telling you the same thing does not add much. And you can pass this information on to someone else all by yourself.

But some things—like norms and behaviors—might not spread this way. They might require a more complex process that involves reinforcement by multiple social contacts. If so, then a network arranged as a simple line, like a bucket brigade, might not support transmission of more complicated phenomena. If we wanted to get people to quit smoking, we would not arrange them in a line and get the first one to quit and tell him to pass it on. Rather, we would surround a smoker with multiple nonsmokers, perhaps in a squad.

Psychologist Stanley Milgram's famous sidewalk experiment illustrates the importance of reinforcement from multiple people.[12] On two cold winter afternoons in New York City in 1968, Milgram observed the behavior of 1,424 pedestrians as they walked along a fifty-foot length of street. He positioned "stimulus crowds," ranging in size from one to fifteen research assistants, on the sidewalk. On cue, these artificial crowds would stop and look up at a window on the sixth floor of a nearby building for precisely one minute. There was nothing interesting in the window, just another guy working for Milgram. The results were filmed, and assistants later counted the number of people who stopped or looked where the stimulus crowd was looking. While 4 percent of the pedestrians stopped alongside a "crowd" composed of a single individual looking up, 40 percent stopped when there were fifteen people in the stimulus crowd. Evidently, the decisions of passersby to copy a behavior were influenced by the size of the crowd exhibiting it.

An even larger percentage of pedestrians copied the behavior incompletely: they looked up in the direction of the stimulus crowd's gaze but did not stop. While one person influenced 42 percent of passersby to look up, 86 percent of the passersby looked up if fifteen people were looking up. More interesting than this difference, however, was that a stimulus crowd of five people was able to induce almost as many passersby to look up as fifteen people did. That is, in this setting, crowds larger than five did not have much more of an effect on the actions of passing individuals.

RULE 5: THE NETWORK HAS A LIFE OF ITS OWN

Social networks can have properties and functions that are neither controlled nor even perceived by the people within them. These properties can be understood only by studying the whole group and its structure, not by studying isolated individuals. Simple examples include traffic jams and stampedes. You cannot understand a traffic jam by interrogating one person fuming at the wheel of his car, even though his immobile automobile contributes to the problem. Complex examples include the notion of culture, or, as we shall see, the fact that groups of interconnected people can exhibit complicated, shared behaviors without explicit coordination or awareness.

Many of the simple examples can be understood best if we completely ignore the will and cognition of the individuals involved and treat people as if they were "zero-intelligence agents." Consider the human waves at sporting events that first gained worldwide notice during the 1986 World Cup in Mexico. In this phenomenon, originally called *La Ola* ("the wave"), sequential groups of spectators leap to their feet and raise their arms, then quickly drop back to a seated position. The effect is quite dramatic. A group of physicists who usually study waves on the surface of liquids were sufficiently intrigued that they decided to study a collection of filmed examples of *La Ola* in enormous soccer stadiums; they noticed that these waves usually rolled in a clockwise direction and consistently moved at a speed of twenty "seats per second."[13]

To understand how such human waves start and propagate, the scientists employed mathematical models of excitable media that are ordinarily used to understand inanimate phenomena such as the spread of a fire through a forest or the spread of an electrical signal through cardiac muscle. An *excitable medium* is one that flips from one state to another (like a tree that is either on fire or not) depending on what others around it are doing (are nearby trees on fire?). And these models yielded accurate predictions of the social phenomenon, suggesting that *La Ola* could be understood even if we knew nothing about the biology or psychology of humans. Indeed, the wave cannot be understood by studying the actions of a single individual standing up and sitting down. It is not orchestrated by someone with a megaphone atop a cooler. It has a life of its own.

Mathematical models of flocks of birds and schools of fish and swarms of insects that move in unison demonstrate the same point: there is no central control of the movement of the group, but the group manifests a kind of collective intelligence that helps all within it to flee or deter predators. This behavior does not reside within individual creatures but, rather, is a property of groups. Examination of flocks of birds "deciding" where to fly reveals that they move in a way that accounts for the intentions of all the birds, and, even more important, the direction of movement is usually the best choice for the flock. Each bird contributes a bit, and the flock's collective choice is better than an individual bird's would be.[14] Similar to *La Ola* and to flocking birds, social networks obey rules of their own, rules that are distinct from the people who form them. But now, people are not having fun in a stadium: they are donating organs or gaining weight or feeling happy.

In this regard, we say that social networks have emergent properties. *Emergent properties* are new attributes of a whole that arise from the interaction and interconnection of the parts. The idea of emergence can be understood with an analogy: A cake has a taste not found in any one of its ingredients. Nor is its taste simply the average of the ingredients' flavors—something, say, halfway between flour and eggs. It is much more than that. The taste of a cake transcends the simple sum of its ingredients. Likewise, understanding social networks allows us to understand how indeed, in the case of humans, the whole comes to be greater than the sum of its parts.

Six Degrees of Separation and Three Degrees of Influence

Stanley Milgram masterminded another, much more famous experiment showing that people are all connected to one another by an average of "six degrees of separation" (your friend is one degree from you, your friend's friend is two degrees, and so on). Milgram's experiment, conducted in the 1960s, involved giving a few hundred people who lived in Nebraska a letter addressed to a businessman in Boston, more than a thousand miles away.[15] They were asked to send the letter to somebody they knew personally. The goal was to get it to someone they thought would be more likely than they to have a personal relationship with the Boston businessman. And the number of hops from person to person that the letter took to reach the target was tracked. On average, six hops were required. This amazing fact initiated a whole set of investigations into the small-world effect originally characterized by de Sola Pool and Kochen, and it entered popular culture too, with John Guare's play *Six Degrees of Separation* and even the trivia game Six Degrees of Kevin Bacon.

But some academics were skeptical. For instance, as far apart as Nebraska and Boston might be (both geographically and culturally), they were both inside the United States. So in 2002, physicist-turned-sociologist Duncan Watts and his colleagues Peter Dodds and Roby Muhamad decided to replicate Milgram's experiment on a global scale using e-mail as the mode by which people communicated.[16] They recruited more than ninety-eight thousand subjects (mostly from the United States) to send a message to "targets" around the world by forwarding the e-mail to someone each subject knew who might in turn know the targeted person. Each subject was randomly assigned one target from a list of eighteen possible targets in thirteen countries. The targets included a professor at an Ivy League university, an archival inspector in Estonia, a technology consultant in India, a policeman in Australia, and a veterinarian in the Norwegian army—quite a motley crew. Once again—astonishingly—it took roughly six steps (on average) to get the e-mail to each targeted person, replicating Milgram's original estimate of just how small the world is.

However, just because we are connected to everyone else by six degrees of separation does not mean that we hold sway over all of these people at any social distance away from us. Our own research has shown that the spread of influence in social networks obeys what we call the Three Degrees of Influence Rule. Everything we do or say tends to ripple through our network, having an impact on our friends (one degree), our friends' friends (two degrees), and even our friends' friends' friends (three degrees). Our influence gradually dissipates and ceases to have a noticeable effect on people beyond the social frontier that lies at three degrees of separation. Likewise, we are influenced by friends within three degrees but generally not by those beyond.

The Three Degrees Rule applies to a broad range of attitudes, feelings, and behaviors, and it applies to the spread of phenomena as diverse as political views, weight gain, and happiness. Other scholars have documented that among networks of inventors, innovative ideas seem to diffuse to three degrees, so that an inventor's creativity influences his colleagues, his colleagues' colleagues, and his colleagues' colleagues' colleagues. And word-of-mouth recommendations for everyday concerns (like how to find a good piano teacher or how to find a home for a pet) tend to spread three degrees too.

There are three possible reasons our influence is limited. First, like little waves spreading out from a stone dropped into a still pond, the influence we have on others may eventually peter out. The stone displaces a certain volume of water as it is dropped, and the energy in the wave dissipates as it spreads out. One way to think about this socially is that there is decay in the fidelity of information as it is transmitted, as in the child's game of telephone. So, if you quit smoking or endorse a particular political candidate, by the time this information reaches your friends' friends' friends' friend, that person may no longer have accurate or reliable information about what you actually did. We call this the *intrinsic-decay explanation.*

Second, influence may decline because of an unavoidable evolution in the network that makes the links beyond three degrees unstable. Ties in networks do not last forever. Friends stop being friends. Neighbors move. Spouses divorce. People die. The only way to lose a direct connection to someone you know is if the tie between you disappears. But for a person three degrees removed from you, any of three ties could be cut and you would lose at least one pathway between you. Hence, on average, we may not have stable ties to people at four degrees of separation given the constant turnover in ties all along the way. Consequently, we do not influence nor are we influenced by people at four degrees and beyond. We call this the *network-instability explanation.*

Third, evolutionary biology may play a part. As we will discuss in chapter 7, humans appear to have evolved in small groups in which everyone would have been connected to everyone else by three degrees or less. It is indeed useful to know whether anyone in our group has it in for us or is our ally, or whether others need our help or might help us. And it is useful to influence others in our group to do what we do. But we have not lived in large groups long enough for evolution to have favored people who can extend their influence beyond three degrees. Put another way, we may not be able to influence people four degrees removed from us because, in our hominid past, there was no one who was four degrees removed from us. We call this the *evolutionary-purpose explanation.*

It seems likely that all these factors play a role. But no matter the reasons, the Three Degrees Rule appears to be an important part of the way human social networks function, and it may continue to constrain our ability to connect, even though technology gives us access to so many more people.

While this inherent limit may seem, well, limiting (who doesn't want to rule the world?), we should remember how small the world is. If we are connected to everyone else by six degrees and we can influence them up to three degrees, then one way to think about ourselves is that each of us can reach about halfway to everyone else on the planet.

Moreover, even when restricted to three degrees, the extent of our effect on others is extraordinary. The way natural social networks are structured means that most of us are connected to thousands of people. For example, suppose you have twenty social contacts, including five friends, five coworkers, and ten family members, and each of them in turn has similar numbers of friends and family (to make things simple, let's assume they are not the same contacts as yours). That means you are indirectly connected to four hundred people at two degrees of separation. And your influence doesn't stop there; it goes one more step to the twenty friends and family of each of those people, yielding a total of 20 × 20 × 20 people, or eight thousand people who are three degrees removed from you. That would include every single person in the small Oklahoma town where James grew up.

So while the observation that there are six degrees of separation between any two people applies to how connected we are, the observation that there are three degrees of influence applies to how contagious we are. These properties, connection and contagion, are the structure and function of social networks. They are the anatomy and physiology of the human superorganism.

Connected

Most of us are already aware of the direct effect we have on our friends and family; our actions can make them happy or sad, healthy or sick, even rich or poor. But we rarely consider that everything we think, feel, do, or say can spread far beyond the people we know. Conversely, our friends and family serve as conduits for us to be influenced by hundreds or even thousands of other people. In a kind of social chain reaction, we can be deeply affected by events we do not witness that happen to people we do not know. It is as if we can feel the pulse of the social world around us and respond to its persistent rhythms. As part of a social network, we transcend ourselves, for good or ill, and become a part of something much larger. We are connected.

Our connectedness carries with it radical implications for the way we understand the human condition. Social networks have value precisely because they can help us to achieve what we could not achieve on our own. In the next few chapters, we will show how networks influence the spread of joy, the search for sexual partners, the maintenance of health, the functioning of markets, and the struggle for democracy. Yet, social-network effects are not always positive. Depression, obesity, sexually transmitted diseases, financial panic, violence, and even suicide also spread. Social networks, it turns out, tend to magnify whatever they are seeded with.

Partly for this reason, social networks are creative. And what these networks create does not belong to any one individual—it is shared by all those in the network. In this way, a social network is like a commonly owned forest: we all stand to benefit from it, but we also must work together to ensure it remains healthy and productive. This means that social networks require tending, by individuals, by groups, and by institutions. While social networks are fundamentally and distinctively human, and ubiquitous, they should not be taken for granted.

If you are happier or richer or healthier than others, it may have a lot to do with where you happen to be in the network, even if you cannot discern your own location. And it may have a lot to do with the overall structure of the network, even if you cannot control that structure at all. And in some cases, the process feeds back to the network itself. A person with many friends may become rich and then attract even more friends. This rich-get-richer dynamic means social networks can dramatically reinforce two different kinds of inequality in our society: *situational inequality* (some are better off socioeconomically) and *positional inequality* (some are better off in terms of where they are located in the network).

Lawmakers have not yet considered the consequences of positional inequality. Still, understanding the way we are connected is an essential step in creating a more just society and in implementing public policies affecting everything from public health to the economy. We might be better off vaccinating centrally located individuals rather than weak individuals. We might be better off persuading friends of smokers of the dangers of smoking rather than targeting smokers. We might be better off helping interconnected groups of people to avoid criminal behavior rather than preventing or punishing crimes one at a time.

The powerful effect of social networks on individual behaviors and outcomes suggests that people do not have complete control over their own choices. Interpersonal influence in social networks therefore raises moral questions. Our connections to others affect our capacity for free will. How much blame does Giacomo in Corsica deserve for his actions, and how much credit does Dan Lavis in Ontario deserve for his? If they acted merely as links in a chain, how can we understand their freedom to choose their actions at all?

Some scholars explain collective human behavior by studying the choices and actions of individuals. Others dispense with individuals and focus exclusively on groups formed by social class, race, or political party affiliation, each with collective identities that cause people in these groups to mysteriously and magically act in concert. The science of social networks provides a distinct way of seeing the world because it is about individuals *and* groups, and about how the former actually become the latter.

If we want to understand how society works, we need to fill in the missing links between individuals. We need to understand how interconnections and interactions between people give rise to wholly new aspects of human experience that are not present in the individuals themselves. If we do not understand social networks, we cannot hope to fully understand either ourselves or the world we inhabit.

CHAPTER 2

When You Smile,
the World Smiles with You

A strange thing happened in Tanzania in 1962. At a mission boarding school for girls near Lake Victoria in the Bukoba District, there was an epidemic of laughter. And this was not just a few schoolgirls sharing a joke. An irresistible desire to laugh broke out and spread from person to person until more than one thousand people were affected.

The affliction had an abrupt onset, and the initial bout of laughter lasted between a few minutes and a few hours in those affected. This was followed by a period of normal behavior, then typically a few relapses over the course of up to sixteen days. In what was to be a clue about the real nature of this epidemic, the victims often described feeling restless and fearful, despite their laughter.

The physicians who first investigated and reported on the outbreak—Dr. Rankin, a faculty member at Makerere University, and Dr. Philip, the medical officer of the Bukoba District—were extremely thorough.[1] They found that each new patient had recent contact with another person suffering from the malady. They were able to observe that the incubation period between contact and onset of symptoms ranged from a few hours to a few days. Thankfully, as they intoned without irony, "no fatal cases have been reported." Afflicted persons recovered fully.

The epidemic began on January 30, 1962, when three girls aged twelve to eighteen started laughing uncontrollably. It spread rapidly, and soon most people at the school had a serious case of the giggles. By March 18, ninety-five of the 159 pupils were affected, and the school was forced to close. The pupils went home to their villages and towns. Ten days later, the uncontrollable laughter broke out in the village of Nshamba, fifty-five miles away, where some of the students had gone. A total of 217 people were affected. Other girls returned to their village near the Ramanshenye Girls' Middle School, and the epidemic spread to this school in mid-June. It too was forced to close when forty-eight of 154 students were stricken with uncontrollable laughter. Another outbreak occurred in the village of Kanyangereka on June 18, again when a girl went home. The outbreak started with her immediate family and spread to two nearby boys' schools, and those schools were also forced to close. After a few months, the epidemic petered out.

Rankin and Philip looked hard for biological causes for the epidemic. They performed physical examinations and lab studies on the patients, did spinal taps, examined the food supply for toxins, and ascertained that there was no prior record of a similar epidemic in the region. The villagers themselves did not know what to make of it. In Bukoba, where the illness aroused great interest, there was the "belief that the atmosphere had been poisoned as a result of the atom bomb explosions." Others described it as a kind of "spreading madness" or "*endwara yokusheka*," which means simply, "the illness of laughing."

As the villagers and the scientists investigating this outbreak realized, the epidemic was no laughing matter. It did not involve the spread of real happiness and joy—though this can happen too, albeit not in quite the same way. Rather, the outbreak was a case of epidemic hysteria, a condition that takes advantage of a deep-rooted tendency of human beings to exhibit emotional contagion. Emotions of all sorts, joyful or otherwise, can spread between pairs of people and among larger groups. Consequently, emotions have a collective and not just an individual origin. How you feel depends on how those to whom you are closely and distantly connected feel.

Our Ancestors Had Feelings

We all have emotions. And they consist of several elements. First, we usually have a conscious awareness of our emotions: when we are happy, we know it. Second, emotions typically affect our physical state: we show how we feel on our faces, in our voices, even in our posture; given the role emotions play in social networks, these physical manifestations are especially important. Third, emotions are associated with specific neurophysiological activity; if you are shown a scary picture, the flow of blood to structures deep in your brain instantly changes. Finally, emotions are associated with visible behaviors, like laughing, crying, or shrieking.[2]

Experiments have demonstrated that people can "catch" emotional states they observe in others over time frames ranging from seconds to weeks.[3] When college freshmen are randomly assigned to live with mildly depressed roommates, they become increasingly depressed over a three-month period.[4] Emotional contagion can even take place between strangers, after just ephemeral contact. When waiters are trained to provide "service with a smile," their customers report feeling more satisfied, and they leave better tips.[5] People's emotions and moods are affected by the emotional states of the people they interact with. Why and how does this happen?

We might consider another question first: Why aren't emotions merely internal states? Why don't we just have our own private feelings? Having feelings is surely evolutionarily advantageous to us.

For example, the ability to feel startled is probably good for us in situations where we need to react quickly to survive. But we do not just feel startled, we show that we are startled. We jump or shriek or curse or clench, and these actions do not go unnoticed. They are copied by others.

Given the organization of early hominids into social groups, the spread of emotions served an evolutionarily adaptive purpose.[6] Early humans had to rely on one another for survival. Their interactions with the physical environment (weather, landscape, predators) were modulated or affected by their interactions with their social environment. Humans bonded with others in order to face the world more effectively, and mechanisms evolved to support this bonding, most obviously verbal communication but also emotional mimicry. The development of emotions in humans, the display of emotions, and the ability to read the emotions of others helped coordinate group activity by three means: facilitating interpersonal bonds, synchronizing behavior, and communicating information.

Emotions and emotional contagion probably first arose to facilitate mother-infant pair bonding and then evolved to extend to kin members and ultimately to nonkin members. Emotional contagion fosters interaction synchrony. At the level of mother-child pairs, emotional contagion may have prompted mothers to be more attentive to and protective of their babies when their babies needed attention. Indeed, we are sadder when our family members are sad than when strangers are sad. There is an advantage in coordinating our moods with those to whom we are related.

Eventually this type of synchrony in mood or activity may have been beneficial for larger group activities, such as warding off enemies or hunting prey. If you are trying to coordinate a hunting party, it helps if members of the group are all upbeat and fired up. Conversely, if you are part of a group and someone in it appears afraid, perhaps that person has seen a predator that you have not seen. Quickly adopting his emotional state can enhance your prospects for survival. Indeed, it is thought that positive emotions may work especially well to increase group cohesiveness ("I'm happy; stay with me") and that negative emotions may work well as communication devices ("I smell smoke; I'm scared").

Emotions may be a quicker way to convey information about the environment and its relative safety or danger than other forms of communication, and it seems certain that emotions preceded language. What emotions lack in specificity compared to oral language, they may make up for in speed. You can tell whether your spouse is mad at you very quickly, but having her explain it to you may take a good deal more time (especially if she insists that you guess why she is mad before she tells you). You can walk through the door at home at the end of the day and immediately know whether the environment is safe or dangerous, and that is quite a trick our ancestors bequeathed us.

Of course, rapidly coordinated emotions are not always a good thing. If you come home and are in a bad mood, your partner will often detect it long before you resort to the more laborious process of explaining why you are in a bad mood. And before you have a chance to explain, she might already have caught your bad mood, which may lead to an argument and a downward spiral.

Emotional Contagion

Emotions spread from person to person because of two features of human interaction: we are biologically hardwired to mimic others outwardly, and in mimicking their outward displays, we come to adopt their inward states. If your friend feels happy, she smiles, you smile, and in the act of smiling you also come to feel happy. In bars and bedrooms, at work and on the street, everywhere people interact, we tend to synchronize our facial expressions, vocalizations, and postures unconsciously and rapidly, and as a result we also meld our emotional states.

Nowhere do we show our emotions more than on our faces. It is not difficult to explain why our facial expressions change in response to environmental stimuli or how this may be evolutionarily adaptive. Recent research, for example, has provided insight into how two facial expressions, fear and disgust, moderate our reception of sensations coming from the outside world.[7] When we are terrified, our eyes widen and our nostrils flare to help us see and smell more of our surroundings, just as the ears of a dog perk up when it hears something interesting. Similarly, when we are disgusted, such as by an offensive odor, our noses wrinkle and our eyes narrow to reduce the impact. Air intake increases when we are afraid and decreases when we are disgusted.

Yet, facial expressions appear to have evolved not just to modify our experience of the world as individuals but as a way to communicate with others. Over time, this aspect of facial expressions probably eclipsed their original role. Such changes happen often in evolution. Feathers may have arisen merely to insulate the bodies of prehistoric reptiles, but they wound up contributing to a different and more important advantage, the ability to fly.

We developed an ability to read the facial expressions of others. Hence, we benefit when our own faces are contorted in disgust *and* by being able to notice whether others' faces are contorted in disgust. Humans have an extraordinary knack for detecting even small changes in facial expressions. This ability is localized in a particular area of the brain and can even be lost, a condition tongue-twistingly known as *prosopagnosia*. Reading the expressions of others was probably a key step on the way toward synchronizing feelings and developing the emotional empathy that underlies the process of emotional contagion.

Even as early as 1759, it was apparent to founding economist and philosopher Adam Smith that conscious thought was one way we could feel for others and hence feel like others: "Though our brother is upon the rack…by the imagination we place ourselves in his situation, we conceive ourselves enduring all the same torments, we enter as it were into his body, and become in some measure the same person with him, and thence form some idea of his sensations, and even feel something which, though weaker in degree, is not altogether unlike them."[8]

However, emotions spread in ways beyond simply reading faces and thinking about the experiences of others. There is actually a more primitive, less deliberative process of emotional contagion, a kind of instinctive empathy. People imitate the facial expressions of others, then, as a direct result, they come to feel as others do. This is called *affective afference,* or the facial-feedback theory, since the path of the signals is from the muscles (of the face) to the brain, rather than the more usual, efferent pathway from the brain to the muscles. The beneficial effects of facial expressions on a person's mood are among the reasons, for example, that telephone operators are trained to smile when they work, even though the person at the other end of the line cannot see them. This theory also explains why it helps to smile when your heart is breaking.

One biological mechanism that makes emotions (and behaviors) contagious may be the so-called *mirror neuron system* in the human brain.[9] Our brains practice doing actions we merely observe in others, as if we were doing them ourselves. If you've ever watched an intense fan at a game, you know what we are talking about—he twitches at every mistake, aching to give his own motor actions to the players on the field. When we see players run, jump, or kick, it is not only our visual cortex or even the part of our brain that thinks about what we are observing that is activated, but also the parts of our brain that would be activated if we ourselves were running, jumping, or kicking.

In one experiment related to emotional contagion, subjects listened to recordings of nonverbal vocal reactions communicating two positive emotions, such as amusement and triumph, and two negative emotions, such as fear and disgust. Investigators monitored the subjects' brains for a response by placing them in a magnetic resonance imaging (MRI) machine.[10] The subjects were told not to react to what they heard. While subjects did not visibly respond to the sounds, the MRI results showed that hearing the cues stimulated parts of the brain that command the corresponding facial expressions. It seems we are always poised to feel what others feel and to do what others do.

Emotional Stampedes

Everyone has experience with emotional contagion: we share a joke with a friend, we feel sad when a spouse cries, we rage against city hall with our neighbors, and we hug our kids tight when they've had a bad day. Yet one often overlooked aspect of all this sharing is that emotions spread not only to our friends but to our friends' friends and beyond—even when we are not present. We are like a herd of buffalo quietly grazing on the plain until one of our neighbors starts to run. Then we start to run, and others start to run, and suddenly, mysteriously, the whole herd is barreling forward.

Epidemics of emotional states have been reported for centuries. They just have not involved laughter like the Bukoba outbreak. When emotions spread from person to person and affect large numbers of people, it is now called *mass psychogenic illness* (MPI) rather than the old-fashioned and more poetic epidemic hysteria. MPI is a specifically social phenomenon involving otherwise healthy people in a psychological cascade. Like a single startled buffalo within a herd, a single emotional reaction in one person can sometimes cause many others to feel the same thing, creating an emotional stampede.

There are two main types of MPI. In the *pure-anxiety type,* people may feel a variety of physical symptoms, including abdominal pain, headache, fainting, shortness of breath, nausea, dizziness, and so on. In the *motor type,* people may engage in hysterical dancing, pseudo-seizures, and—yes—laughing, though the actual feelings underlying these behaviors are fear or anxiety. Both types of MPI thus involve the same basic psychological processes.

Historical records of such phenomena date back to at least 1374, when, in close succession to the Black Death in Europe, "dancing manias" broke out. The first such manias occurred in what is now Aachen, Germany. As described by the German medical historian J. F. C. Hecker in his 1844 book *The Epidemics of the Middle Ages,* these consisted of people who "united by one common delusion, exhibited to the public both in the streets and in the churches the following strange spectacle. They formed circles hand in hand, and appeared to have lost all control over their senses, continued dancing, regardless of the bystanders, for hours together, in wild delirium, until at length they fell to the ground in a state of exhaustion. They then complained of extreme oppression, and groaned as if in the agonies of death."[11] These people were obviously no happier to be dancing than the African schoolgirls were to be laughing.

In a bygone era, demons and witchcraft were often seen as causes of these symptoms, but today toxic chemicals and environmental contamination are the triggers subjects typically identify. Yet, while toxins do cause some outbreaks of physical illness, they do not cause outbreaks of MPI. The source of the problem, as well as the mechanism of transmission, is psychological. Individuals afflicted in these outbreaks, and many observers, are often reluctant to ascribe the symptoms to a psychological source, however.

A relatively recent example of MPI occurred at the Warren County High School in McMinnville, Tennessee. At the time, the school had 1,825 students and 140 staff members. On November 12, 1998, a teacher believed she smelled gasoline, which caused her to complain of headache, shortness of breath, dizziness, and nausea. Seeing her response, some of her students soon developed similar symptoms. As the classroom was being emptied, other students, observing what was happening, began to report feeling unwell too. A schoolwide fire alarm was activated, and the school was evacuated.

The teacher and several students were transported by ambulance to a nearby hospital, in full view of other students and teachers who were outside because of the alarm. Large numbers of police, firefighters, and emergency medical personnel from three counties responded. A total of one hundred people went to the hospital that day, and thirty-eight were admitted. Classes were canceled.

The school was closed for four days. It was inspected by the fire department, the gas company, and state officials from the Occupational Safety and Health Administration (OSHA), but no problems were identified. After the school had been deemed perfectly safe, the students and the teachers were allowed to return. Unfortunately many still smelled odors, and on November 17, seventy-one people were stricken. Ambulances were again called, and the school was evacuated and then closed.

The school's principal was fed up. In a "no more Mr. Nice Guy" move, he decided to call several government agencies, including the famed Epidemic Intelligence Service of the Centers for Disease Control (CDC). Also involved were the federal Environmental Protection Agency, the Agency for Toxic Substances and Disease Registry, the National Institute for Occupational Safety and Health, OSHA, the Tennessee Department of Health, the Tennessee Department of Agriculture, and numerous other local emergency organizations and personnel. The investigation was extremely thorough. Aerial surveillance identified potential environmental sources of contamination; personnel explored caves in the vicinity of the school; the school's air-handling, plumbing, and structural systems were thoroughly checked; core samples were drilled from the grounds around the school; and air samples (including from the days of the outbreak) and water and waste samples were tested. The air was evaluated with an astonishing array of technology, including colorimetric tubes, flame-ionization detectors, photoionization detectors, radiation meters, and combustible-gas indicators.

Two years later, a *New England Journal of Medicine* article described the extensive examination of possible environmental causes for the illness and reported the results of the investigation by the CDC. In the end, like Rankin and Philip studying the African laughter epidemic, the investigators concluded that psychogenic factors were to blame. They found that the illness was associated with directly observing another ill person during the outbreak and with being female.[12] The diagnosis was epidemic hysteria.

This diagnosis did not sit well with the community, and it upset many of those who had been ill, such as one twelfth grader who was quoted as saying, "They said we were crazy....It just made me mad. When I'm sick, I don't want someone to say I'm faking. They wouldn't have taken me to the hospital, and my blood pressure wouldn't have been sky-high, if I wasn't sick."[13] Of course, the symptoms of those with MPI, whether laughing, dancing, fainting, or nausea, are quite real; they do not "fake" their experience in the deliberate, premeditated way that a malingerer does. The astonishing reality is that our own anxiety makes us sick, but so does the anxiety of others.

The CDC investigators also discussed why communities tended to use so many resources to try to find environmental causes for conditions that appeared to be psychogenic. The problem is that while public health professionals often suspect that an outbreak is psychogenic, they feel they have no choice but to conduct an unreasonably thorough investigation because of intense anxiety in the community. And, of course, it is very difficult, if not impossible, to definitively prove that a mysterious toxic exposure has not simply escaped detection. The CDC investigators noted the possibility of a negative community reaction to an episode labeled as psychogenic, saying, "Physicians and others are understandably reluctant to announce that an outbreak of illness is psychogenic because of the shame and anger that the diagnosis tends to elicit."[14]

An Unbearable Sweetness

Outbreaks of epidemic hysteria are not restricted to children and schools. They have been documented in adults too. One systematic review of cases of epidemic hysteria identified seventy outbreaks that occurred between 1973 and 1993 and found that 50 percent of them took place in schools, 40 percent in small towns and factories, and only 10 percent in other settings.[15] The outbreaks usually involved at least thirty people, and often hundreds. Most outbreaks lasted less than two weeks, but 20 percent lasted more than a month.

One of the more improbable examples was the case of the "phantom anesthetist of Mattoon." In 1944, over a period of a few weeks during the climax of World War II, many adult residents of Mattoon, Illinois, became convinced that an "evil genius" was on the loose in their town of fifteen thousand people. This unseen person would open bedroom windows and spray victims with a "sweet-smelling" anesthetic gas that would temporarily paralyze them but, strangely, leave others in the same room unaffected. Citizens banded together to form armed patrols, but the anesthetist was never caught. The local sheriff, fearing that an innocent person might be shot, eventually ordered the posses to disband. As one investigator of this outbreak dryly noted, "The 'gasser' hypothesis asserts that the symptoms were produced by a gas which was sprayed on the victims by some ingenious fiend who has been able to elude the police. This explanation...is widely believed in Mattoon at present. The alternative hypothesis is that the symptoms were due to hysteria."[16]

Another, more recent case occurred in 1990 among the Triborough Bridge toll employees in New York City. On February 16, workers began to complain of headaches, abdominal discomfort, dizziness, and throat and chest pain. More and more workers came down with the same symptoms over the next several days, with some of the ill workers noting what they described as a "sweetness" in the air. Symptoms were reported when workers were inside or near a toll booth, but they would subside soon after workers left the booths. The outbreak ended on February 22, when some of the workers' superiors sat with them at the tolls. By that time, thirty-four workers had become ill enough to go to the hospital, and many others shared their symptoms. After spending hundreds of thousands of dollars searching in vain among dozens of potential culprits for a physical cause of the symptoms, it became clear to many that the illness was psychogenic. It forced 44 percent of the female workers to go to the hospital, almost twice the proportion of male workers with debilitating symptoms.

These cases share many characteristics of MPI. The symptoms tend to pop up in and spread through highly connected communities (with high network transitivity). These communities tend to be isolated and stressed. A physical culprit is seldom found. In most cases, the majority of those affected are women. It is not clear why the incidence in women and girls is higher, but it is possible that because women are inclined to discuss their symptoms, more sympathy cases result in other women. The fact that women have a more sensitive sense of smell might also play a role.

For some reason that is not well understood, smells, both real and imagined, are frequent triggers of modern outbreaks of MPI. This may have to do with the well-established connection between olfaction and emotions. Experiments have demonstrated that smell and emotion are both regulated by a part of the brain called the orbitofrontal cortex.[17] Experiments have also shown that memories evoked by smell induce stronger emotions than those evoked by verbal descriptions of the same odor.[18] Words are powerful, but one familiar whiff can jolt the mind into the past with more emotional intensity than can a signal from any other sense. This is called the *Proust phenomenon,* after the author who described a poignant memory inspired by the scent of a cookie. Smelling a perfume associated with a happy memory leads to more activity in the amygdala (a part of the brain involved in emotion and emotional memory) than seeing the bottle that the perfume comes in.[19]

Paradoxically, the presence of official personnel—whether police officers, rescue workers, scientific investigators, or government officials—often worsens the epidemic, for it reinforces the belief that something serious is going on and that the situation is potentially dangerous. When these same officials attempt to provide reassurance that the situation is safe and that no cause was found, it typically generates deep suspicions among the emotionally charged populace that a cover-up is under way, especially because the official response was previously so substantial. Paranoia can spread too, undermining the very authority that is needed to bring an end to such an episode.

The recommended treatment for MPI outbreaks focuses on social networks and recognizes that social ties are the medium for spread. The psychological guidelines for emergency workers include "providing reassurance...using a calm and authoritative approach" and "separating those who are ill from those who are not."[20] As one expert put it, "You can only stop these things by being honest....I could get caught up in this kind of thing too, as a parent or just a person. We all could. It's a very powerful thing, and it needs to be respected and understood. And health officials shouldn't be so scared to call a spade a spade."[21]

It's often difficult to establish why exactly these epidemics start. Just as an unfamiliar noise can trigger a cattle herd to start running, many triggers can cause emotional stampedes. However, it is usually fairly simple to identify the initial cases. For example, in the African laughing epidemic, even though the investigators could not explain why it started, they easily located the first girls to have symptoms.

It only took a few people to start *La Ola* in the stadium in Mexico City or to get passersby to stop and look up at a window in New York City, and the same is true of MPI outbreaks. When a small group of people begin acting in concert or experiencing similar, visible symptoms, the epidemic can spread along social-network ties via emotional contagion, and large groups can very quickly become emotionally synchronized.

The present obsession with nut allergies in the United States may be a case in point. The number of schools declaring themselves to be entirely "nut free" is by all accounts rising. Nuts and staples like peanut butter are prohibited from campus, and so are homemade baked goods and any foods without detailed ingredient labels. School entrances have signs admonishing visitors to wash their hands before entering to safeguard students from possible contamination.

Approximately 3.3 million Americans are allergic to nuts, and even more, 6.9 million, are allergic to seafood. However, all told, serious allergic reactions to foods cause just two thousand hospitalizations per year (out of more than thirty million hospitalizations nationwide). And, at most, only 150 people (both children and adults) die each year from food allergies. Compare that to the fifty people who die each year from bee stings, the hundred who die from lightning strikes, and the forty-five thousand who die from motor vehicle accidents. Or compare that to the ten thousand children who are hospitalized each year for traumatic brain injuries acquired during sports, or the two thousand who drown, or the roughly thirteen hundred who die from gun accidents. Yet there are no calls to end athletics. There are likely thousands of parents who rid their cupboards of peanut butter but not guns. And more children assuredly die walking or being driven to school each year than die of nut allergies.

The question is not whether nut allergies exist, or whether they can occasionally be serious, or whether reasonable accommodations should be made for the few children who have documented serious allergies. The question is, what accounts for society's extreme response to nut allergies? Not surprisingly, the response bears many of the hallmarks of MPI. A few people have clinically documented concerns, but others who do not then copy the behaviors of those who do. Anxiety spreads from person to person to person, and a sense of proportion and the ability to be reassured are lost.

Well-intentioned efforts to reduce nut exposure actually fan the flames since they indicate to parents that nuts are a clear and present danger. This encourages more parents to worry, which fuels the epidemic. It also encourages more parents to have their kids tested, thus detecting mild and meaningless allergies to nuts. And, finally, this encourages still more avoidance of nuts, which may actually lead to a rise in true nut allergies because lack of exposure to allergens early in life is thought to contribute to the onset of allergies later.[22]

MPI is a pathological phenomenon, but it takes advantage of a nonpathological process that is fundamental in humans, namely, the tendency to mimic the emotional state of others. Real laughter also can be contagious and so can real happiness. But comparing epidemic hysteria to these more normal processes is like comparing the stampede of a herd to its more usual and orderly migration.

Tracking the Spread of Emotions

Measuring the subjective experience of emotions (as compared with their visible, biological, or neurological manifestations) requires asking people how they are feeling. One of the more systematic ways of doing this is known as the *experience-sampling method.* This method uses a series of alerts (such as signals sent to a beeper or cell phone) at unexpected times to prompt subjects to document their feelings, thoughts, and actions while they are experiencing them.[23] The result is a thorough picture of the ups and downs of subjects' daily emotional lives.

One of the advantages of this method is that it allows groups of interacting people to be evaluated simultaneously in real time. For example, one team of investigators, interested in the spread of emotions within families, outfitted fifty-five families (consisting of a mother, father, and one adolescent) with beepers for one week. The participants were beeped roughly every 90 to 120 minutes between 7:30 a.m. and 9:30 p.m., and a total of 7,100 time points were observed in these 165 individuals.

Various emotional states were measured, such as whether the subjects were happy or unhappy. Although the investigators could not rule out the possibility that the entire family was simultaneously exposed to one thing that made them all sad or happy at once (a confounding effect that we will discuss in greater detail in chapter 4), they did try to tease out how emotions spread within these families.

The strongest path was from daughters to both parents, while, conversely, the parents' emotional state appeared to have no effect on their daughters. Fathers' emotions affected their wives and their sons but not their daughters. This appeared to be especially true when fathers returned from work: when dad came home in a lousy mood, he soon made the whole household miserable.[24]

A similar method has been used to examine the transmission of emotions among teams of nurses, athletes, and even accountants.[25] In such professional settings, a key question was whether one fired-up team member could improve the mood and thus the performance of his teammates. Not surprisingly, positive mood is associated with a range of team-performance-enhancing changes, including greater altruistic behavior, increased creativity, and more efficient decision making. A nice demonstration involved outfitting thirty-three professional male cricket players with pocket computers that recorded their moods four times a day during a match (which can have the insane duration of five days). There was a strong association between a player's own happiness and the happiness of his teammates, independent of the state of the game; further, when a player's teammates were happier, the team's performance improved.

The Spread of Happiness

Despite the biological and psychological evidence for emotional mimicry, and the numerous cases of MPI arising from epidemic anxiety, until recently little was known about the precise role of social networks in the spread of emotions. Yet, the MPI cases suggest that emotions spread far and wide, flowing through social-network ties from person to person to person, and that there should be a normal analogue to this pathological phenomenon. Indeed, there can be waves of emotions in the vast fabric of human social relationships, so that people in particular locations in the social network have one emotional experience, and others elsewhere who come under different influences have a different experience altogether.

Strangely, while researchers in diverse fields, including medicine, economics, psychology, neuroscience, and evolutionary biology, have identified a broad range of stimuli of individual human happiness, they have not addressed a key (perhaps *the* key) determinant: the happiness of others. It may be obvious that our friends and family can make us happy, but before we undertook our own investigation, no one had ever explored how happiness can spread through social networks from person to person to person.

We became curious about this. We were particularly interested in determining whether the spread of emotions occurred not just between you and your friends (dyadic spread) but also between you and your friends' friends, and their friends, and beyond (hyperdyadic spread). How far did emotions travel in the network? And were there geographic or temporal constraints on the spread?

Our first step in answering these questions was to assemble a data set that had measures of emotions and social connections over time. (We discuss that process in chapter 4.) We then created a graph of the social network of happiness, as shown in plate 1. This illustration shows ties among siblings, friends, and spouses in a sample drawn from 12,067 people originally from Framingham, Massachusetts, in the year 2000, along with their levels of happiness. No one had ever plotted such a graph before. One thousand twenty people are represented, and each node is colored on a spectrum from blue (unhappy) to yellow (happy) according to the subject's level of happiness. Looking at this image suggests two observations. First, unhappy people cluster with unhappy people in the network, and happy people cluster with happy people. Second, unhappy people seem more peripheral: they are much more likely to appear at the end of a chain of social relationships or at the edge of the network.[26]

Clustering of this kind in social networks can arise from a variety of processes. Happy people might choose each other as friends or be exposed to the same environments that cause them all to be happy at the same time. But our analyses allowed us to adjust for these effects. And we found that clustering is also due to the causal effect of one person's happiness on another's. Mathematical analyses of the network suggest that a person is about 15 percent more likely to be happy if a directly connected person (at one degree of separation) is happy. And the spread of happiness doesn't stop there. The happiness effect for people at two degrees of separation (the friend of a friend) is 10 percent, and for people at three degrees of separation (the friend of a friend of a friend), it is about 6 percent. At four degrees of separation, the effect peters out. Here we have our first evidence of the Three Degrees of Influence Rule. Emotions (and, as we will see later, norms and behaviors) spread in social networks from person to person to person, but they do not spread to everyone. Just as a ripple in a pond eventually fades away, so too does the ripple of an individual's happiness fade through the social network.

At first glance, these effects may not seem very significant. But compare them to the effect of having a higher income. An extra $5,000 in 1984 dollars (which corresponds to about $10,000 in 2009 dollars) was associated with only a 2 percent increased chance of a person being happy. So, having happy friends and relatives appears to be a more effective predictor of happiness than earning more money. And the amazing thing is that even people who are three degrees removed from you, whom you may have never met, can have a stronger impact on your personal happiness than a wad of hundreds in your pocket. Being in a particular spot in a social network, exposed to people with particular feelings, has important implications for your life.

It is well known that having more friends and relatives is much more likely to put a smile on your face than having more cash.[27] But past research had never considered why friends matter so much. There are at least two possibilities. First, the existence of the social relationship itself may improve your happiness—this is a structural effect of the network on you (the second rule of social networks described in chapter 1). As we discuss in chapter 7, we are hardwired to seek out social relationships, so it is not surprising that we feel pleasure or reward when we spend time with friends and family. Second, friends and relatives make us susceptible to emotional contagion, so our friends' emotional states affect our own (the third rule of social networks).

While both of these mechanisms probably contribute to people's happiness, our evidence suggests that contagion may be the more important of the two. We found that each happy friend a person has increases that person's probability of being happy by about 9 percent. Each unhappy friend decreases it by 7 percent. So if you were simply playing the averages, and you didn't know anything about the emotional state of a new person you just met, you would probably want to be friends with her. She might make you unhappy, but there is a better chance she will make you happy. This helps to explain why past researchers have found an association between happiness and the number of friends and family. But once we control for the emotional states in one's friends, we find that having more friends is not enough—having more happy friends is the key to our own emotional well-being.

This does not mean that the structure of the social network is unimportant. Amazingly, it is not just the number of dyadic ties that has an impact; the number of hyperdyadic ties also influences a person's happiness. When we measured the centrality of each person in the social network, we found that people with more friends of friends were also more likely to be happy. And, more remarkably, this was true even among people who had the same number of direct social relationships. This means that the more friends your friends have (regardless of their emotional state), the more likely you are to be happy.

One might wonder if there is a chicken-and-egg problem here. After all, it is possible to imagine that when we become happier, we tend to attract more friends, and more friends who have lots of friends. This would mean that happiness is driving the network rather than the other way around. But when we examined how the network changes over time, we found that happy people do not tend to become more central. So having a wide social circle can make you happy, but being happy does not necessarily widen your social circle. Being located in the middle of the network leads to happiness rather than the other way around. The structure of your network and your location in it matter.

Given how important direct interaction seems to be for emotional contagion to occur, we also theorized that the effect of the happiness of your social contacts on your emotional state should depend on how near or far they are. The idea is that people who live nearby are more likely to be in contact and therefore more likely to pick up on each others' moods. Geographic distance can be used as a proxy for frequency of social interaction. In our study, about one in three people live within a mile of their closest friend, but there is a lot of variation, and some friends live thousands of miles apart. We found that when a friend who lives less than a mile away becomes happy, it can increase the probability that you are happy by 25 percent. In contrast, the happiness of a friend who lives more than a mile away has no effect. Similarly, if your spouse lives with you and he or she becomes happy, then your probability of happiness goes up, but spouses who do not live together (because they are separated) have no effect on each other. A happy sibling who lives less than a mile away increases your chance of happiness by 14 percent, but more distant siblings have no significant effect. And happy next-door neighbors also increase your chance for happiness, while neighbors who live farther away (even on the same block) have no significant effect.

All these findings suggest the importance of proximity among people whose emotions influence each other, and the impact of immediate neighbors suggests that the spread of happiness may depend as much on frequent face-to-face interaction as on deep personal connections. While in this case we are considering the spread of a dispositional state of some duration, these findings are also in keeping with the work on facial mimicry we discussed earlier.

Happiness is thus not merely a function of individual experience or choice; it is also a property of groups of people. Changes in individual happiness can ripple through social connections and create large-scale patterns in the network, giving rise to clusters of happy and unhappy individuals. Since our work was published, similar results on the spread of happiness have been observed in a sample of ten thousand rural Chinese villagers.[28] Although we could not observe what causes happiness to spread, a variety of mechanisms are conceivable. Happy people may share their good fortune (e.g., by being pragmatically helpful or financially generous to others), change their behavior toward others (e.g., by being nicer or less hostile), or merely exude an emotion that is contagious. Being surrounded by happy people might have beneficial biological effects too. But whatever the mechanism, it seems clear that we need to change the way we think about happiness and other emotions.

Life on the Hedonic Treadmill

We all know people who are hedonists; they can never get enough of the good life. In fact, lasting happiness is difficult to achieve because people are on a "hedonic treadmill." Although a change in a person's circumstances may cause him to be happier (e.g., finding a partner, winning the lottery) or sadder (e.g., losing a job, becoming paralyzed), a broad body of research has shown that people tend to return to their previous level of happiness after such events.[29] In fact, studies of lottery winners and spinal cord injury patients reveal that after a year or two, they are often no more happy or sad than the rest of us. Our surprise at this observation stems in part from our inability to anticipate that some things will not change. Lottery winners still have annoying relatives, and paralyzed patients can still fall in love. As psychologist Daniel Gilbert has shown, we tend to focus on only the most salient part of a situation when we are thinking about things that might befall us.[30] Moreover, we overlook our ability to adapt to circumstances. So, a person trying to become happier is like someone walking up a downward-moving escalator. Although the effort to climb up and become happier is helpful, it is counteracted by the process of adaptation that forces one back to one's original state.

Many people try to overcome this problem by intentionally engaging in activities to improve their happiness. We might change our behavior by exercising regularly or by trying to be kind to others or even by avoiding a long commute (which has been shown to be particularly deleterious to happiness). We might change our attitude by pausing to count our blessings or thinking about experiences in the most positive light (as Tibetan monks do). We might also devote effort to causes we find meaningful or strive to achieve important personal goals. Indeed, there is reason to suspect that a sustained effort to engage in such happiness-producing activities might help us progress up the downward-moving escalator.

But in spite of these efforts, each of us tends to stay put in a particular long-term disposition; we appear to have a set point for personal happiness that is not easy to change. In fact, like other personality traits, personal happiness appears to be strongly influenced by our genes. Studies of identical and fraternal twins show that identical twins are significantly more likely to exhibit the same level of happiness than are fraternal twins or other siblings. Behavior geneticists have used these studies to estimate just how much genes matter, and their best guess is that long-term happiness depends 50 percent on a person's genetic set point, 10 percent on their circumstances (e.g., where they live, how rich they are, how healthy they are), and 40 percent on what they choose to think and do.[31] What we experience in life can, of course, change our moods for a period of time, but in most cases these changes are transitory.

What about the network spread of happiness? Does it obey this constraint, only making us happy for a while? Does the effect of having a friend become happy tend to wear off? In our study, we found that a person is 45 percent more likely to be happy if a friend became happy in the previous six months. In contrast, the effect is only 35 percent for friends who became happy within the previous year, and it disappears after longer periods of time. So, our friends' happiness does have an effect on us, but it only lasts for about a year. Just as lottery winners get used to their newfound wealth, we get used to our friends being happy. But if different friends get happy at different points in time, they might give us a periodic boost, helping us to stay above our natural level of happiness.

Alone in the Crowd

If happiness can spread, at least for a while, what about other emotions? One feeling that directly concerns our social network is loneliness. In some sense, loneliness is the opposite of connection—it is the feeling of being disconnected. Work by psychologist John Cacioppo has shown that loneliness is a complex set of feelings experienced by people whose core needs for intimacy and social connection are not met.[32] This often motivates most (but not all) people to redress their situation, suggesting that the function of loneliness is to promote reconnection (we will discuss the evolutionary purpose of loneliness in chapter 7).

Psychologists have identified the way that feelings of loneliness fit in with a broad set of other feelings and states, including self-esteem, anxiety, anger, sadness, optimism, and shyness. Psychological research suggests that feelings of loneliness occur when there is a discrepancy between our desire for connection to others and the actual connections we have. This research has focused on the subjective perception of being alone, but feeling lonely is not the same thing. While studies have shown, unsurprisingly, that having a good friend can decrease loneliness, what has not previously been examined is the effect of the whole social network on our tendency to feel lonely even in a crowd.

Using the same network in which we studied happiness, we examined whether being alone was associated with feeling lonely and whether such feelings could spread.[33] We found that real-world social connections do have an effect on how we feel. People with more friends are less likely to experience loneliness. Each extra friend reduces by about two days the number of days we feel lonely each year. Since on average (in our data) people feel lonely forty-eight days per year, having a couple of extra friends makes you about 10 percent less lonely than other people. Interestingly, the number of family members has no effect at all. It is not clear why this is the case. Possibly, people in small families know they have a greater responsibility to spend time with one another since there are fewer people to take turns visiting. Or perhaps people in large families primarily feel close to a smaller core of their family, limiting the influence of additional connections. Regardless of the mechanism, it is clear that feelings of loneliness are much more closely tied to our networks of optional social connections than to those handed to us at birth.

Loneliness can actually shape the social network. People who feel lonely all the time will lose about 8 percent of their friends, on average, over two to four years. Lonely people tend to attract fewer friends, but they also tend to name fewer people as friends as well. What this means is that loneliness is both a cause and a consequence of becoming disconnected. Emotions and networks can reinforce each other and create a rich-get-richer cycle that rewards those with the most friends. People with few friends are more likely to become lonely, and this feeling then makes it less likely that they will attract or try to form new social ties.

Our study suggests that physical proximity matters as much for loneliness as it does for happiness. Friends and family who live nearby see each other more often, which should help decrease the likelihood that they feel lonely, but it also makes them more susceptible to one another's feelings. For example, if a nearby friend has ten extra lonely days a year, it will increase the number of lonely days you experience by about three. If this person is a close friend, then the effect is stronger, and you'll experience four extra days of loneliness. Loneliness also spreads between next-door neighbors, with ten extra days of loneliness leading to two extra days for the person on the other side of the fence. But neighbors and friends who live more than a mile away do not make each other lonely.

Spouses who live together can affect each other too, but the result is less dramatic. For every ten extra days a person is lonely, his or her spouse will be lonely for just one extra day. And siblings do not appear to affect one another at all (even the ones who live nearby); this provides additional evidence that loneliness is about our relationships to people with whom we choose to connect rather than the relationships we have inherited.

Looking beyond these direct connections, we found that loneliness spreads three degrees, just like happiness. A person's loneliness depends not only on his friends' loneliness, but also on his friends' friends' and his friends' friends' friends' loneliness. The full network shows that you are about 52 percent more likely to be lonely if a person you are directly connected to (at one degree of separation) is lonely. The effect for people at two degrees of separation is 25 percent, and for people at three degrees of separation, it is about 15 percent. At four degrees of separation the effect disappears, in keeping with the Three Degrees of Influence Rule.

Finally, we observed an extraordinary pattern at the edge of the social network. At the periphery, people have fewer friends; this makes them lonely, but this also tends to drive them to cut the few ties that they have left. But before they do, they may infect their friends with the same feeling of loneliness, starting the cycle anew. These reinforcing effects mean that our social fabric can fray at the edges, like a strand of yarn that comes loose from the sleeve of a sweater. If we are concerned about combating the feeling of loneliness in our society, we should aggressively target the people at the periphery with interventions to repair their social networks. By helping them, we can create a protective barrier against loneliness that will keep the whole network from unraveling.

Feeling in Love

The psychology of emotions such as happiness and loneliness sheds light on the formation and dissolution of ties in social networks. In fact, human sensibilities such as anger, sadness, grief, and love all operate in the service of social ties. One can be angry at nature or saddened by a forest fire or love a pet, but these emotions have their origin and find their fullest expression in the anger, sadness, or love one feels in the setting of interpersonal relationships.

People the world over have different ideas, beliefs, and opinions—different thoughts—but they have very similar, if not identical, feelings. And they have similar responses to feelings in others, preferring happy friends to depressed ones, kind friends to mean ones, and loving friends to violent ones. A whole range of emotions can spread, from anger and hatred to anxiety and fear to happiness and loneliness. But there is one emotion central to human experience that we have not yet considered and that is key to understanding social connection: love.

The psychology of love and affection is obviously crucial to an understanding of the formation of social ties between people. As anthropologist Helen Fisher has argued, the sensibility of being in love may be broken down into lust, love, and attachment, all of which likely served evolutionary purposes.[34] The feeling of lust has the obvious goal of driving reproduction—with almost any partner. The feeling of romantic love is something different, of course, and tends to be focused on a particular partner, or at least one partner at a time. From an evolutionary perspective, this allows the individual to conserve precious resources and not waste them in the pursuit of several objects of affection. The feeling of attachment, and the secure tie to another person that it represents, may have evolved to allow parents to jointly care for their young, which also has evolutionary advantages.

In chapter 7, we will discuss the role of natural selection in human social networks in more detail, but before we get there it is important to think about the implications of our deepest connections. Aside from the evolutionary advantages and disadvantages, feelings of lust, love, and attachment carry enormous implications for the way we connect to others. The object of one's affection becomes the "center of one's universe," around which all else revolves. People experience intrusive thoughts about their beloved, aggrandize their beloved, are energized by their beloved, and are obviously deeply connected to their beloved. We usually experience such romantic love with just one person at a time. So romantic love does not determine the general organization of social networks. After all, we do not love everyone we know. And the love we have for our parents, children, siblings, and other relations is a different kind of feeling. Yet, as we will see in the next chapter, being in love is a key mechanism by which certain important social ties are formed, and it is therefore highly relevant to the origin—and function—of social networks.

CHAPTER 3

Love the One You're With

N icholas and his wife, Erika, like to joke that they had an arranged marriage, South Asia style. Though they lived within four blocks of each other for two years and were both students at Harvard, their paths never crossed. Erika had to go all the way to Bangladesh so that Nicholas could find her. In the summer of 1987, he went to Washington, DC, where he had grown up and gone to high school, to care for his ailing mother. He was a medical student, single, and, he foolishly thought, not ready for a serious relationship. His old high-school friend, Nasi, was also home for the summer. Nasi's girlfriend, Bemy, who had come to know Nicholas well enough that her gentle teasing was a source of amusement for all of them, was also there. She had, as it turned out, just returned from a year in rural Bangladesh, doing community development work.

In the waterlogged village where Bemy had spent her year abroad was a beautiful young American woman with whom she shared a burning desire to end poverty and a metal bucket to wash her hair.

You probably know where this story is going. One afternoon, in the middle of the monsoon, while writing a postcard to Nasi, Bemy suddenly turned to her friend Erika and blurted out: "I just thought of the man you're going to marry." That man was Nicholas. Erika was incredulous. But months later, she agreed to meet him in DC, when the four of them had dinner at Nasi's house. Nicholas was of course immediately smitten. Erika was "not unimpressed," as she later put it. That night, after getting home, Erika woke up her sister to announce that she had, indeed, met the man she was going to marry. Three dates later, Nicholas told Erika he was in love. And that is how he came to marry a woman who was three degrees removed from him all along, who had practically lived next door, who had never known him before but who was just perfect for him.

Such stories — with varying degrees of complexity and romance — occur all the time in our society. In fact, a simple Google search for "how I met my wife" and "how I met my husband" turns up thousands of narratives, lovingly preserved on the Internet. They can be short, such as this one: "How did I meet my husband? At a bar. He was a friend of the scummy boyfriend, soon-to-be-husband of my best friend (yes, they're divorced). I was introduced to him in a bar...hooked up...and we're still together, and married...while my best friend isn't!"

Or the stories can be more involved: "I drove into the valley of Yosemite National Park sometime after the sun went down with my two girlfriends and a pitbull. I had worked there the two summers before and was preparing for another season. When we stepped out of the car, it was freezing, and we had to trudge through a foot of snow up to our friend's cabin. He wasn't home but had left a note directing us to another cabin. We were wet up to our calves by the time we reached it, and I felt uncomfortable knocking on a stranger's door. Luckily, our friend opened up and invited us in to his friend's cabin. He made introductions, and I must've seemed rude because

I ran to the heater and turned my back to the room. Somehow, the occupancy level diminished without me realizing it, and I ended up sitting on a bed opposite my future husband. He reminded me of a young Dave Matthews. His southern accent was charming, and those eyes…God, those eyes. We talked well into the night until my friend, who had settled into a bed near me, sighed and begged for us to leave. I thanked him for having us, and he said, 'Well, now you know where I live so drop in anytime.' Back in the cold Sierra night, we giggled all the way down to the parking lot where I turned to my girlfriends and said those fateful words, 'I'm going to marry that man!' Two years and five months later, I did."[1]

How I Met My Partner

The romantic essence of these stories is that they seem to involve both luck and destiny. But, if you think about it, these meetings aren't so chancy. What these stories really have in common is that the future partners started out with two or three degrees of separation between them before the gap was inexorably closed.

The romantic ideal of finding a partner often also involves the sense that you have the right "chemistry" with your intended or that the two of you fall in love for mysterious, inexplicable reasons. We think of falling in love as something deeply personal and hard to explain. Indeed, most Americans believe that their choice of a partner is really no one else's business. Some people select their partners impulsively and spontaneously; others quite deliberately. Either way, partner choice is typically seen as a personal decision. This view of relationships is consistent with our general tendency to see major life decisions as individual choices. We like to believe that we are at the helm of our ship, charting an entirely new course, no matter how choppy the seas. It's surprising and maybe even disappointing to discover that we are in fact sailing through well-traveled shipping lanes using universal navigational tools.

Because we are so sure of our individual power to make decisions, we lose sight of the extraordinary degree to which our choice of a partner is determined by our surroundings and, in particular, by our social network. This also helps to explain the romantic appeal of stories involving putatively chance encounters, for they seem to suggest that forces larger than ourselves are at work, and that romance with a particular, unknown person is predestined and magical. Now, we are not suggesting there isn't something amazing about meeting the love of your life after trudging through the snow at Yosemite or washing your hair in a bucket in Bangladesh. It's just that those magical moments are not as random as we might think.

Consider some systematic data about how people meet their partners. The National Survey of Health and Social Life, also quaintly known as the Chicago Sex Survey, studied a national sample of 3,432 people aged eighteen to fifty-nine in 1992 and provides one of the most complete and accurate descriptions of romantic and sexual behavior in the United States.[2] It contains detailed information about partner choice, sexual practices, psychological traits, health measures, and so on. It also includes a type of data that is surprisingly very rare, namely, how and where people actually met their current sexual partners. The table shows who introduced couples in different kinds of relationships.

Who introduced the couple?

Type of relationship	Close relationships		More-distant relationships					
	Family member	Friend	Coworker	Classmate	Neighbor	Self-introduction	Other introduction	Number of subjects
Marriages	15%	35%	6%	6%	1%	32%	2%	1,287
Cohabitations	12%	40%	4%	1%	1%	36%	3%	319
Partnerships	8%	36%	6%	4%	1%	42%	1%	920
Short-term partnerships	3%	37%	3%	4%	2%	47%	2%	251

Note: Numbers do not add to 100 percent because of rounding.

The introducers here did not necessarily intend for the two people they introduced to become partners, but the introduction nevertheless had this effect. Roughly 68 percent of the people in the study met their spouses after being introduced by someone they knew, while only 32 percent met via "self-introduction." Even for short-term sexual partners like one-night stands, 53 percent were introduced by someone else. So while chance encounters between strangers do happen, and while people sometimes find their partners without assistance, the majority of people find spouses and partners by meeting friends of friends and other people to whom they are loosely connected.

While friends were a source of introduction for all kinds of sexual partnerships at roughly the same rate (35–40%), family members were much more likely to introduce people to their future spouses than to future one-night stands. And how people meet is also relevant to how quickly they have sex. In the Chicago study, those who met their partners through their friends were slightly more likely to have sex within a month of meeting than those who met through family members. A similar study conducted in France found that couples who met at a nightclub were much more likely to have sex within a month (45 percent) than those who met at, say, a family gathering (24 percent), which is not surprising since one typically does not have sex in mind at family events.[3]

These data suggest that people might use different strategies to find partners for different kinds of relationships. Maybe people ask family members for introductions to possible marriage partners and rely on their own resources to meet short-term partners. This makes intuitive sense: most drunken college students are not texting their mothers to see if they should invite that cute stranger at the bar home for the night. So, what you get when searching your network depends in part on where you are looking and what you are looking for.

However, it is clear that people rely heavily on friends and family for all kinds of relationships. When you meet a new person on your own, you only have information about yourself. In contrast, when others introduce you to someone new, they have information about both you and your potential partner, and sometimes they play the role of matchmaker (consciously or not) by encouraging meetings between people they think will get along. Friends and family are likely to know your personalities, social backgrounds, and job histories, and they also know important details such as your tendency to leave underwear on the floor or to send roses. The socially brokered introduction is less risky and more informative than going it alone, which is one reason people have relied on introductions for thousands of years.

Yet in most modern societies, we generally have a negative view of arranged marriages, and we cannot possibly imagine what it would be like to marry a stranger. Well-meaning friends and relatives who nosily interfere in our lives to help us find partners are seen as comic figures, like Yente in *Fiddler on the Roof*. But, in fact, our friends, relatives, and coworkers typically take on a matchmaking role only when they think we are having trouble finding a partner on our own. And as it turns out, our social network functions quite efficiently as matchmaker, even when we insist we are acting out our own private destiny.

The structure of naturally occurring social networks is perfectly suited to generate lots of leads. In networks such as bucket brigades and phone trees, there are only a limited number of people within a few degrees of separation from any one person. But in most natural social networks, there are thousands. As we discussed in chapter 1, if you know twenty people (well enough that they would invite you to a party), and each of them knows twenty other people, and so on, then you are connected to eight thousand people who are three degrees away. If you are single, one of all these people is likely to be your future spouse.

Of course, random encounters can sometimes bring strangers together, especially when incidental physical contact is involved.

These happy accidents are frequently used as plot devices in romantic stories, whether it's two people grabbing the same pair of gloves in *Serendipity*, an umbrella taken by mistake after a concert in *Howard's End*, or dogs getting their leashes entangled in *101 Dalmatians*. Incidents like these provide opportunities for further social interaction, and possibly sex or marriage, because they require what sociologist Erving Goffman called "corrective" rituals: people have to undo the "damage," and this in turn means that they have to get to know each other. Good flirts are able to turn such happenstance into real opportunities. And the best flirts may even be able to contrive an "accident" in order to meet someone: they make their own luck. But these are the exceptions more than the rule. And it is noteworthy that even these meetings of strangers involve some degree of shared interest, whether in clothing, music, or pets, for instance.

Even when people meet on their own, without help from mutual contacts, there is a social preselection process that influences the kinds of people they are likely to run into in the first place. For example, the Chicago Sex Survey also collected data on where Americans met their partners. Sixty percent of the people in the study met their spouses at places like school, work, a private party, church, or a social club—all of which tend to involve people who share characteristics. Ten percent met their spouses at a bar, through a personal ad, or at a vacation spot, where there is more diversity but still a limited range of types of people who might be available to become future spouses.[4]

The locations and circumstances under which people meet partners have been changing over the past century. Our best data on this come from a study conducted in France. Looking across a broad range of venues where people meet spouses, including nightclubs, parties, schools, workplaces, holiday destinations, family gatherings, or simply "in the neighborhood," the investigators traced the change across time. For example, from 1914 until 1960, 15 to 20 percent of people reported meeting their future spouses in the neighborhood, but by 1984 this percentage was down to 3 percent, reflecting the decline of geographically based social ties as a consequence of modernity and urbanization.[5]

Geography is even less important with the rise of the Internet. In 2006, one in nine American Internet-using adults—all told, about sixteen million people—reported using an online dating website or other site (such as Match.com, eHarmony.com, or the wonderfully named PlentyofFish.com, as well as countless others) to meet people.[6] Of these "online daters," 43 percent—or nearly seven million adults— have gone on actual, real-life dates with people they met online, and 17 percent of them—nearly three million adults—have entered long-term relationships or married their online dating partners, according to a systematic national survey.[7] Conversely, 3 percent of Internet users who are married or in long-term committed relationships reported meeting their partners online, a number that will likely rise in the coming years.[8] Gone are the days of the girl next door. People increasingly meet their partners through (offline and online) social networks that are much less constrained by geography than they used to be.

My Partner Is Just Like Me

With the decline in importance of meeting people in the neighborhood in recent years, people no longer search geographic space for partners. Nevertheless, they still search social space. Rather than going from house to house or town to town, we jump from person to person in search of the perfect mate. We see if anyone near us in our network (e.g., our friends, coworkers) would be a suitable partner, and if not, we look farther away (e.g., our friends' friends, our coworkers' siblings). And we often seek out circumstances, such as parties, that are likely to result in meeting friends of friends and people still farther away in our network.

We have "weak ties" to friends of friends and other sorts of people we do not know very well. But, as we will discuss in chapter 5, these kinds of ties can be incredibly valuable for connecting us to people we do not know at all, thereby giving us a much greater pool of people to choose from. So the best way to search your network is to look beyond your direct connections but not so far away that you no longer have anything in common with your contacts. A friend's friend or a friend's friend's friend may be just the person to introduce you to your future spouse.

Some societies have richly prescriptive procedures for partner search, and although they severely limit personal choice for the betrothed, they still exploit network connections. Such marriages are often arranged for legal or economic reasons rather than from a desire to find a suitable partner (in the Western sense), and they are common in the Middle East and Asia. In some cultural settings, customs prescribe that the prospective partners be introduced to each other, and the parents take an active role in vetting the family and the potential spouse. In other settings, however, the marriage is a settled matter from the first meeting, and no courtship is allowed. Across cultures, there is considerable variation in who the matchmakers are (parents, professionals, elders, clergy), what pressures the matchmakers can exert, what qualifications the spouses must have (reputation, wealth, caste, religion), and what sanctions can be imposed if the couple refuses (disinheritance, death).

These practices are not immutable, however, even in societies where arranged marriage was formerly the norm. For example, the percentage of women living in Chengdu in Sichuan, China, who had arranged marriages shrank from 68 percent of those married between 1933 and 1948 to 2 percent of those wed between 1977 and 1987.[9] Nevertheless, social-network ties still remain crucial; 74 percent of respondents in Chengdu report that the primary network that connects young people to potential mates is friends and relatives in the same age group.

Regardless of what kind of network people use, whether real

or virtual, the process of searching for a mate is usually driven by *homogamy*, or the tendency of like to marry like (just as *homophily* is the tendency of like to befriend like). People search for—or, in any case, find—partners they resemble (in terms of their attributes) and partners who are of comparable "quality." The Chicago Sex Survey, for example, shows that the great majority of marriages exhibit homogamy on virtually all measured traits, ranging from age to education to ethnicity. Other studies show that spouses usually have the same health behaviors (like eating and smoking), the same level of attractiveness, and the same basic political ideology and partisan affiliation (with rare, notable exceptions like Clinton adviser James Carville and Republican strategist Mary Matalin). We would expect more homophily in long-term relationships and less in short-term relationships (one is less finicky when it comes to sexual partners than potential spouses), and to some extent this is indeed the case: 72 percent of marriages exhibit homophily (based on a summary measure involving several traits), compared to 53 to 60 percent for other types of sexual relationships.[10] In addition, as we shall see, spouses also become more similar over time because they influence each other (for example, in political affiliation, smoking behavior, or happiness).

On the one hand, homogamy makes intuitive sense. People like being around others who are similar to them. Most people find it comforting to imagine that partners resemble each other because it gives them hope that they, too, will someday be happy in a warm and loving relationship with a kindred spirit. On the other hand, think about the odds of finding someone just like you. Personal ads are full of complex laundry lists that must be very difficult, if not impossible, to satisfy: *Wanted: frisky, down-to-earth, nonsmoking, leftist Democrat salsa dancer who likes guns, Bollywood films, NASCAR races, Ouija boards, beach sunsets, Cosmopolitans, country drives, and triathlons.*

Indeed, the uniqueness of each human being has implications for how many people out there are a perfect fit in this sense. The age-old debate about whether you have one soul mate or a million rests in part on how picky you are. But even if there are a million compatible people for you, that is just one of every six thousand people in the whole world. If you are choosing at random, you had better go on a lot of dates. The dispiritingly unromantic conclusion is that you will never, ever find Mr. or Ms. Right. Not without some help.

But the surprising power of social networks is that they bring likes together and serve up soul mates in the same room. Bigger and broader social networks yield more options for partners, facilitate the flow of information about suitable partners via friends and friends of friends, and provide for easier (more efficient, more accurate) searching. Hence, they yield "better" partners or spouses in the end. The odds of finding that soul mate just improved substantially.

Given the structure of social networks, our tendency to be introduced to our partners, and our innate comfort with people we resemble, it is not surprising that we generally wind up meeting, having sex with, and marrying people like ourselves. The choice of a partner is constrained by the same social forces that create network ties in the first place. Who we befriend, where we go to school, where we work—all these choices largely depend on our position in a given social network. No matter where people search, their network generally acts to bring similar people together. The fact that spouses are so often similar manifestly disproves the idea that people meet and choose their partners by chance.

Big Fish, Little Pond

American satirist H. L. Mencken famously observed that wealth is "any income that is at least one hundred dollars more per year than the income of one's wife's sister's husband." With this statement, he captured an idea that is well known to most people but strangely unpopular in the formal study of economics: namely, that people often care more about their relative standing in the world than their absolute standing. People are envious. They want what others have, and they want what others want. As economist John Kenneth Galbraith argued in 1958, many consumer demands arise not from innate needs but from social pressures.[11] People assess how well they are doing not so much by how much money they make or how much stuff they consume but, rather, by how much they make and consume compared to other people they know.

An essential truth in Mencken's quip is that the two men are comparing themselves to those from whom they are three degrees removed. They do not compare themselves to strangers. Instead, they seem intent on impressing people they know. In a classic experiment investigating this phenomenon, most people reported that they would rather work at a company where their salary was $33,000 but everyone else earned $30,000 than at another, otherwise identical company where their salary was $35,000 but everyone else earned $38,000.[12] Even though their absolute income is less at the first job, they think they would be happier working there than at the second. We would rather be big fish in a small pond than bigger fish in an ocean filled with whales.

Perhaps not surprisingly, this is also true of our desire to be attractive. In one creative experiment, respondents were asked which of the following two states they would rather be in:

A: Your physical attractiveness is 6; others average 4.
B: Your physical attractiveness is 8; others average 10.

Overall, 75 percent of people preferred being in situation A than in situation B. For most people, their relative attractiveness was more important than their absolute attractiveness.[13] We have repeated this experiment with Harvard undergraduates, and their responses were even more skewed: 93 percent preferred situation A, and 7 percent situation B. And, of course, any bridesmaid forced to wear an unflattering dress understands this point.

These results show that our preference for relative attractiveness is more extreme than our preference for relative income. People realize how crucial it is to have sex appeal if they are to have sex. And they realize how important it is to be more attractive than their prospective mate's other choices. In other words, relative standing is important if it has what is known as an *instrumental payoff*: a more appealing physique than others is a means to an end.

This preference for relative standing brings to mind another classic anecdote: Two friends are hiking in the woods and come to a river. They take off their shoes and clothes and go for a swim. As they come out of the water, they spot a hungry bear that immediately starts to run toward them. One of the men starts fleeing immediately, but the other pauses to put on his shoes. The first man screams at the second, "Why are you putting on your shoes? They won't help you outrun the bear!" To which the second man calmly responds: "I don't need to outrun the bear; I just need to outrun you."

It is this same reasoning that drives ever-larger numbers of people to have plastic surgery and with greater frequency. Liposuction might yield a physical advantage for early adopters, but when everybody gets it done, the advantage goes away. As a result, people then demand other kinds of plastic surgery in a kind of silicone arms race. The breadth of services demanded explodes to parallel the spread of services through the network.

Competition for mates can actually be quite stressful. One investigation we conducted suggests that the higher the male-to-female ratio at a time when a man reaches his early twenties, the shorter his life. A man who is surrounded by other men has to work harder to find a partner, and this environment of elevated competition has long-term consequences for his health. In this regard, we are no different from a number of animal species. In one analysis, we examined the effect of the gender ratio in a sample of high-school seniors in Wisconsin in 1957—a total of 4,183 young men and 5,063 young women in 411 high schools. We found that men in high-school graduating classes with lopsided gender ratios (of more men) wound up with shorter life spans fifty years later. In another analysis of more than 7.6 million men from throughout the United States, we found that the availability of marriageable women again had a durable impact on men's health, affecting their survival well into their later years.[14]

These results suggest that the people who surround us are not only a source of partners or of information about partners; they also are our chief competitors. As a result, the social network in which we find ourselves defines our prospects. It does so by defining whom we meet, by influencing our taste in what is deemed desirable in a partner, and, finally, by specifying how we are perceived by others and what competitive advantages and disadvantages we have. You don't need to be the most beautiful or most wealthy person to get the most desirable partner; you just need to be more attractive than all the other women or men in your network. In short, the networks in which we are embedded function as *reference groups,* which is a social scientist's way of saying "pond."

In the 1950s, Robert K. Merton, a very influential social scientist, codified the basic ways that reference groups affect us: they can have *comparative effects* (how we or others evaluate ourselves), *influence effects* (the way others dictate our behaviors and attitudes), or both.[15] Having unattractive social contacts may make us feel superior (comparison) but may also make us take worse care of ourselves (influence). These two effects may work at cross-purposes in our quest to find a partner.

For decades, reference groups have been seen as abstract categories: people often compare themselves to other "middle-class Americans" or other "members of their grade at school" or other "amateur soccer players." But exciting advances in network science are now enabling us to map out exactly who these references group are for each person. Many people may be more attractive than we are, but our only real competitors are the people in our intended's social network.

Everyone Else Is Doing It

People we know influence how we think and act when it comes to sex. To begin with, both friends and strangers affect our perceptions of a prospective partner's attractiveness, consciously and unconsciously. These effects go beyond basic tendencies that men and women have to make judgments about appearance; for example, it has repeatedly been shown that men find women with low waist-to-hip ratios more attractive, and women value certain facial features in men. Until recently, most research on partner choice and assessments of attractiveness has focused on an individual's independent preferences. Yet there are good biological and social reasons to suppose that perceptions of attractiveness can spread from person to person.

An experiment suggests how. First, investigators took pictures of men who were rated equally attractive by a group of women.[16] Then, they presented pairs of pictures of two equally attractive men to another group of women, but between each pair of pictures, they inserted a picture of a woman who was "looking" at one of the men. This woman was smiling or had a neutral facial expression. The female subjects were much more likely to judge a man to be more attractive than his competitor if the woman interposed between the photos was smiling at him than if she was not.

In another study, a group of women again rated photographs of men for attractiveness. The photos were accompanied by short descriptions, and when the men were described as "married," women's ratings of them went up.[17] In still another study, men in photographs with attractive female "girlfriends" were judged to be more attractive when the "girlfriend" was in the photo than when she was not. Having a plain "girlfriend," however, did not enhance a man's appeal as much.[18] And, astoundingly, women's preferences for men who are already attached may vary according to where the women are in their menstrual cycles. When they are in the fertile phase of the cycle, they have a relative preference for men who are already attached to other women.[19]

There is thus a kind of unconscious social contagion in perceptions of attractiveness from one woman to another. This makes perfect sense from an evolutionary perspective. Copying the preferences of other women may be an efficient strategy for deciding who is a desirable man when there is a cost (in terms of time or energy) in making this assessment or when it is otherwise hard to decide. While a woman can, with a glance, assess for herself various attributes of a man that might be associated with his genetic fitness (his appearance, his height, his dancing ability), other traits related to his suitability as a reproductive partner (his parenting ability, his likelihood of being sweet to his kids) can require more time and effort to evaluate. In those cases, the assessment of another woman can be very helpful. In fact, psychologist Daniel Gilbert has shown that a woman can do a better job of predicting how much she will enjoy a date with a man by asking the previous woman who dated him what he is like than by knowing all about the man.[20] This fact has been exploited for commercial purposes: there is a matchmaking website that only allows men to post if they are "recommended" by a former girlfriend.

In *direct mate choice,* you choose who you like, but in *indirect mate choice* of the sort we have been considering, you choose who others like. Indirect mate choice can even lead people to choose mates with characteristics that they did not previously care about. A slight preference by some women for men with tattoos, for example, can lead hordes of men to get tattoos and inspire other women to want men who have them.

Perhaps not surprisingly, men react differently to social information. While they clearly have shared norms about what is attractive in a woman, contextual cues in men can actually operate in the opposite way.[21] College-age women were more likely to rate a man as attractive if shown a photograph of him surrounded by four women than if shown a photograph of him alone. But college-age men were less likely to rate a woman as attractive if she was shown surrounded by four men than if she was shown alone. This makes evolutionary sense: when selecting mates, males tend to be less choosy than females and so are less concerned with the opinions of anyone else to begin with. But the presence of other men conveys information of a different sort, namely, that there might be time-consuming (and stressful) competition to secure the woman's interest.

Hence, social networks affect our relationships in two important ways. First, structural features of our position in the social network can affect whether people think we are attractive. Do we have a partner already? How connected are we? Do we have many or few partners and friends? Others notice such things about us because they say something about who we are. Second, the social network can spread ideas and change attitudes toward attractiveness. Specific preferences for the opposite sex diffuse, and both men and women come to value partners with certain appearances based on what their friends think. Of course, our friends and families also provide explicit comments on our partners and have a conscious influence on our perceptions and behaviors as well.

Unfortunately, detailed data regarding entire social networks and how sexual attitudes and behaviors spread in networks have been very scarce, and most networks that have been studied over the past century have involved only thirty to three hundred people or so. Recognizing the importance of social networks and anticipating a need for data to study them and their role in sexual behavior and other phenomena (such as youth violence, occupational success, and so on), investigators in North Carolina, including sociologists Peter

Bearman, Richard Udry, Barbara Entwisle, and Kathleen Harris, designed and launched an ongoing, nationwide social-network study of American adolescents in 1994.

Known as the Add Health study, this landmark survey was administered to a whopping 90,118 students in 145 junior-high and high schools all around the United States. About 27,000 of the students and their parents were selected for follow-up surveys in 1994, 1995, and then again in 2001. Hundreds of questions were included on the survey, addressing everything from feelings about friends and family, to participation in church and school clubs, to risky behaviors like taking drugs or engaging in unprotected sex. Each student was asked to identify up to ten friends (five male and five female), most of whom—crucially—were also in the sample. The study also collected information about people's romantic partners. All of this allowed scientists to see for the first time, very large, detailed, and comprehensive social networks and to discern the precise architecture of a person's social ties as it changed across time. We can use these data to identify who is at the center of the network and who is at the periphery, who is located in tightly knit cliques and who prefers to associate with several different groups.

Ties between parents and their adolescent children were critical in the transmission of norms and modeling of behavior. For instance, one study that used the Add Health data showed that girls with a close relationship with their fathers were less likely to become sexually active.[22] However, much more important than parents are the peers in an adolescent's network. Add Health studies have shown that the number of friends, the age and gender of those friends, and their academic performance all affect the onset of sexual activity.[23] Friends' religiosity also affects whether adolescents report having sex, and the effect is strongest in dense social networks, where the adolescents' friends tend to be friends with one another.[24]

What these studies show is that sexual behavior can spread from person to person, and the impact the network has depends on how tightly interconnected people are. But sometimes the story is more complicated. Peter Bearman and his colleague Hannah Brückner explored "virginity pledges," a phenomenon that grew out of a social movement sponsored by the Southern Baptist Church, where teens pledge to abstain from sex, typically until marriage.[25] The initial results, accounting for a range of other influences, showed that pledging substantially and independently reduced the likelihood of sexual debut. However, a much more nuanced picture emerged when the investigators looked at the effect more carefully within the social context of each school.

In a small number of "open" schools, where most opposite-sex friendships and romantic ties occur with individuals outside the school, more pledgers indeed meant delayed sexual debut. Surprisingly, though, in "closed" schools, where most ties occur inside the school, more pledgers meant a greater likelihood of sexual debut. These findings suggest that the pledge movement is an identity movement and not solely about abstaining from sex. In closed schools, adhering to this movement might be beneficial (in terms of delaying sex) when one is in the minority, but if pledging becomes the norm, the psychological benefits of a unique identity are diminished, and the effect is lost. It's not just the pledge itself that constrains behavior; it's whether the pledge confers a unique status. Riding a motorcycle and wearing a black leather jacket emblazoned with a skull and crossbones may give you a special identity in a place where few people own motorcycles, but in a place where everyone rides a motorcycle, it may simply mean you like to save gas.

Of course, peer norms can also increase sexual behavior. In fact, peers are more likely to promote sex than discourage it. Adolescents who believe that their peers would look favorably on being sexually active are more likely to have casual, nonromantic sex.[26] Engaging in oral sex with a partner can actually make one more popular among one's friends.[27] These kinds of peer pressures assuredly underlie the changing mores regarding oral sex seen among American teenagers in the late 1990s. And related studies in adults have shown that people with more partners also have more variety in their sex lives and that they "innovate" more in terms of sexual practices.[28]

Romantic and sexual practices as diverse as contraceptive use, anal sex, fertility decisions, and divorce are all strongly influenced by the existence of these behaviors within one's network. For example, in a paper entitled "Is Having Babies Contagious?" economist Ilyana Kuziemko examined eight thousand American families followed since 1968 and found that the probability that a person will have a child rises substantially in the two years after his or her sibling has a child. The effect is not merely a shift in timing but an increase in the total number of children a person chooses to have.[29] Similar effects have been documented in the developing world, where decisions about how many children to have and whether to use contraception spread across social ties.[30]

We can even understand the increasing acceptability of homosexuality as a social-network process. In 1950, there were probably as many gay people as there are now, but they were by and large deeply closeted. San Francisco politician and gay-rights activist Harvey Milk explicitly pushed his fellow activists to come out to their family members, knowing what effect it would have on the network. As the acceptance of homosexuality gradually increased, more and more people came out of the closet, and so more and more people became aware of gay people in their social network, one or two degrees from them. Uncle Harry, the man next door, the coworker, the friend's friend: all were gay and quite normal and as likely as any heterosexual to be nice to you. This in turn led to a positive-feedback loop, further increasing the acceptability of being gay and the number of people coming out.

Unfortunately, the process can also work in reverse, and stigmatization and discrimination can spread too. The balance in this example and all others we will consider is usually determined by something *external* to the network. Just as a germ and an index case (the first person to get sick) are required to start an epidemic (otherwise, there is nothing happening), something outside the network is often required to get the spread of a new norm, such as tolerance, going—a key issue in social contagion that we will return to in chapter 4.

Dying of a Broken Heart?

Without the tremendous efforts by people like the team at Add Health to collect data about sex and relationships, we would know very little about how sexual practices spread through social networks. In chapter 8, we will explore how the current revolution in network science is being driven in part by the sudden availability of enormous data sets from online sources. It is little wonder, then, that some of the first observations about how connections affect us also coincided with the first efforts to collect data on a society-wide scale in the nineteenth century.

When the British parliament created the Registrar General's Office in 1836 to track births and deaths in England, it did so in order to assure the proper transfer of property rights between generations of the landed gentry. Quite by accident, it also wound up creating fertile ground for the study of human connection. The man appointed to be the first compiler of abstracts in this newly created office was no petty bureaucrat. William Farr was a physician of humble origins and great creativity who used this opportunity to establish the first national vital statistics system in the world. Over the next four decades, he would proceed to analyze these statistics in ways unanticipated by either parliament or the gentry.

Vital statistics were to Farr what the Galapagos finches were to Charles Darwin: an inspiration for a whole new science, and the key to a variety of seminal insights about the human condition. At first, Farr explored the mortality rates of different occupations, the optimal way to classify diseases (his system is still in use today), and the mortality associated with being in different insane asylums. But in 1858, using data from France, Farr discovered something even more notable. His analysis demonstrated that people who were married lived longer than those who were widowed or single.

Farr had inadvertently waded into a debate that had been started in 1749 by the French mathematician Antoine Deparcieux who had investigated the longevity of monks and nuns. Deparcieux claimed that people living in "single blessedness" lived longer than those who were not sequestered and not celibate. In opposition to Deparcieux, other commentators at the time were worried that "the suppression of a physiological function [namely, sex] is prejudicial to health." So, the question was this: is celibacy good for your health, or not?

In his 1858 paper, "Influence of Marriage on the Mortality of the French People," Farr was the first to convincingly answer the question. He was able to document the health benefits of marriage, and, conversely, the adverse health consequences of never marrying or of becoming widowed. As Farr put it, "A remarkable series of observations, extending over the whole of France, enables us to determine for the first time the effect of conjugal condition on the life of a large population." Farr analyzed the data of twenty-five million French adults and concluded: "Marriage is a healthy estate. The single individual is more likely to be wrecked on his voyage than the lives joined together in matrimony."[31] With detailed tables, he showed that, for example, in 1853, among men aged twenty to thirty, there were 11 deaths per 1,000 unmarried men, 7 per 1,000 married men, and 29 per 1,000 widowers. For men aged sixty to seventy, the analogous numbers were 50 per 1,000, 35 per 1,000, and 54 per 1,000.

The story was basically the same for women, although Farr did note that, among women at young ages, being unmarried (and presumably chaste) seemed to prolong their lives. Farr surmised that this probably reflected the increased incidence of death during childbirth for married women—as he put it, the "sorrows of childbearing"—which was very high in that century.

Not long after Farr's observations, other scientists began to speculate about the reasons marriage appeared to extend life. Their explanations, it turns out, are still with us today, though we understand them much better than we did 150 years ago. Figuring out how a connection between two people improves survival helped lay the groundwork for understanding how the connection among many people in complicated social networks affects our health, as we will see in the next chapter. But it also laid the groundwork for social-network science more generally, with respect to a host of phenomena. A couple is the simplest of all possible social networks, and marital health effects illustrate how connection and contagion work.

Some observers in the late nineteenth century argued that marriage merely appeared to offer health benefits; what was really going on, they said, was that married people seemed healthier due to a selection bias. Unhealthy people were less likely to get married, and healthy people were more likely to get married. Writing in 1872, Douwe Lubach, a Dutch physician, argued that those with "physical handicaps, mental sufferings, or infamy" would not marry, thus causing those who did marry merely to appear healthier as a result of marriage.[32] And the mathematician Barend Turksma, writing in 1898, argued that "those with the least vitality, hardly able to provide for themselves, are almost all obliged to spend their life unmarried."[33] That is, the same factors that are responsible for shorter life—whether poverty, mental illness, or other social, mental, or physical limitations—are also responsible for being unable to get married. Consequently, these commentators identified a thorny problem: which came first, health or marriage?

Nineteenth-century observers could not tell. Scientific confusion persisted for a hundred years until the 1960s, when a spate of papers on the topic appeared. A key paper, published in the British medical journal the *Lancet,* was entitled "The Mortality of Widowers," and it again used data from the General Register Office.[34] It analyzed the mortality of 4,486 widowed men for up to five years after the death of their wives, and it did something that Farr could not: it followed the men across time after their wives died and documented when, precisely, the men experienced an increased risk of death. The authors found that the mortality rate was 40 percent higher than expected for the first six months after a spouse's death and then returned to the expected rate shortly thereafter. This spike in mortality has been documented many times since. The close proximity in time of the increased risk of death following a wife's death was the first piece of evidence that supported a causal connection between the death of the wife and the death of the husband. Something about being connected was improving health, and something about losing the connection was worsening health, if only for a while.

Putting aside chance, there are three overarching explanations for this phenomenon. First, like the nineteenth-century scientists, the twentieth-century authors of the *Lancet* paper noted the possibility of homogamy. As these authors noted, homogamy includes "the tendency of the fit to marry the fit, and the unfit the unfit." If unfit people marry each other, we should not be surprised to see that premature death in one spouse would be followed by illness or premature death in the other. They were both unhealthy to begin with.

A second explanation, which began to be seriously addressed in the 1960s, was that there could be a joint unfavorable environment. Maybe the two spouses were both exposed to things that made them more likely to die, such as toxins in the environment or a careening bus. If the bus injured them both but left the husband to survive a bit longer than the wife, we clearly would not say that the wife's death caused the husband's death even if we observed that his death quickly followed hers. This problem is known as *confounding* because this extraneous third factor (the toxins, the bus) confounds the ability of scientists to discern what is really going on.

Third and most important, as Farr himself argued, there might be a true causal relationship between marriage and health. Focusing on widowhood and the health costs of losing a spouse, the authors of the *Lancet* paper correctly pointed out that the death of a man's wife might cause his death, and they provided the quaint illustration that "widowers may become malnourished when they no longer have wives to look after them."

There are a host of biological, psychological, and social mechanisms for such a causal effect of widowhood in both men and women. As the *Lancet* authors noted, "tears, slowed movements, and constipation cannot be the only bodily effects [of widowhood], and whatever may be the other effects, they could scarcely fail to have secondary consequences for resistance to various illnesses." Other scientists writing at this time began to refer to the widowhood effect as "dying of a broken heart," and people actually began to take this metaphor literally, searching for, and ultimately finding, evidence that the risk of heart attack rises immediately after the death of a spouse.[35] Something about being connected to a spouse affects our bodies and our minds.

Amazingly, these three explanations—homogamy, confounding, and a true causal effect—are relevant not only to couples and the struggle to understand whether marriage is salubrious. They are, it turns out, also relevant to other phenomena beyond health and even to the operation of social networks more generally. For example, when we considered the spread of emotions in a family, we had to decide whether happiness really spread or whether grandma brought over a puppy, making everyone in the family happy at once. Or, to pick an economic example: Why might two friends both be poor? Did they befriend each other because they were poor? Did they go into business together and become poor together because the business failed? Or did one become poor and the other follow suit by copying the bad spending habits of the first one?

Why Grooms Gain More

Modern research confirms that marriage is indeed good for you, but the benefits for men and women are different. If we could randomly select ten thousand men to be married to ten thousand women, and if we could then follow these couples for years to see who died when, statistical analysis suggests that what we would find is this: being married adds seven years to a man's life and two years to a woman's life—better benefits than most medical treatments.[36]

Recent innovative work by demographer Lee Lillard and his colleagues, Linda Waite and Constantijn Panis, has focused on untangling how and why this is so. Their research has analyzed what happened to more than eleven thousand men and women as they entered and left marital relationships during the period from 1968 to 1988.[37] The group carefully tracked people before they married until after their marriages ended (either because of death or divorce) and even any remarriages. And they closely examined how marriage might confer health and survival benefits and how these mechanisms might differ for men and women.

The emotional support spouses provide has numerous biological and psychological benefits. Being near a familiar person—even an acquaintance, let alone a spouse—can have effects as diverse as lowering heart rate, improving immune function, and reducing depression.[38] Spouses provide social support to each other and connect each other to the broader social network of friends, neighbors, and relatives. In terms of practical support, the most obvious way that husbands and wives help each other is via the economies of scale derived from a joint household—it costs less to live together than to live apart. Having a spouse is also like having an all-purpose assistant who can at least theoretically meet all kinds of needs. Spouses are reservoirs of information and sources of advice, and hence they influence each other's behavior. They have opinions about everything from whether we should wear our jeans or our seat belts, whether we should eat out or order in, whether we should save more money or blow it all. In part because they have a devoted advocate for their interests, married people choose higher-quality hospitals and are less likely to suffer complications from medical treatment compared to the widowed or the unmarried.[39]

In terms of gender roles, Lillard and Waite found that the main way marriage is helpful to the health of men is by providing them with social support and connection, via their wives, to the broader social world. Equally important, married men abandon what have been called "stupid bachelor tricks."[40] When they get married, men assume adult roles: they get rid of the motorcycle in the garage, stop using illegal drugs, eat regular meals, get a job, come home at a reasonable hour, and start taking their responsibilities more seriously—all of which helps to prolong their lives. This process of social control, with wives modifying their husbands' health behaviors, appears to be crucial to how men's health improves with marriage. Conversely, the main way that marriage improves the health and longevity of women is much simpler: married women are richer.

This cartoonish summary of a large body of demographic research may seem quite sexist and out-of-date. In fact, some demographers have commented that perhaps this is just the age-old story of "trading sex for money": women give men intimacy and a sense of belonging, and men give women cash. It is important to note that these studies involved people who were married in the decades when women had much less economic power than men. But, nevertheless, these results point to something more profound and less contentious, namely, that pairs of individuals exchange all kinds of things that affect their health, and that such exchanges—like any transaction—need not be symmetric either in the type or the amount exchanged.

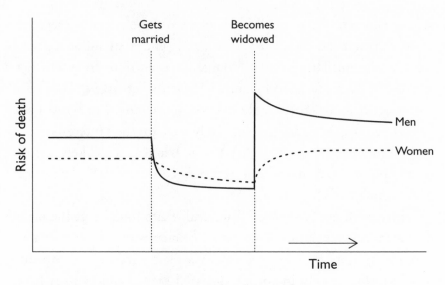

How risk of death changes over time for men and women before, during, and after marriage.

The difference in mechanism across genders and the variation in what is exchanged are also reflected in the timing of health benefits that accrue to men and women when they enter into marriage. When men get married, they experience a sharp and substantial decline in their risk of death (the prompt elimination of stupid bachelor tricks). Women, on the other hand, do not derive an immediate health benefit. It takes longer for them, and their risk of mortality declines more gradually. Moreover, the decline in their risk of death is more modest. These patterns are shown in the illustration.

As we have discussed, something similar happens at widowhood. When a wife dies, the husband's risk of death rises abruptly and dramatically, so that men who lose their wives are between 30 percent and 100 percent more likely to die during the first year of widowhood. This is a clear and persistent finding. Yet, within a few years, widowed men's risk of death declines from this peak.

There has been a lot of debate about whether there is a widowhood effect for women. After Farr's seminal research and until the 1970s, many analyses concluded that women did not suffer any widowhood effect. Then researchers started publishing results that were all over the place, some suggesting that women did not suffer a widowhood effect, and some that they did but to a lesser extent than men. Recent work has concluded that both men and women suffer a widowhood effect, and that it may even be comparable in size.[41]

Questions remain, however, about gender differences in other aspects of the widowhood effect. For example, women might recover sooner from the shock of widowhood than men. Why might there be any discrepancy in magnitude, duration, or mechanism of the effect? Do men suffer more health consequences than women when their spouses die because men love their wives more than women love their husbands? No. Rather, it may be that when men die, the thing they brought to the marriage that had the greatest impact on their spouse's health, namely money, is still around—in the form of assets such as a house and a pension. Conversely, when women die, the thing they brought to a marriage that most affected their partner's health, namely, emotional support, a connection to others, and a well-run home, disappears. Widowed men often find themselves cut off from the social world and lacking social support. Since men in most societies have tended to cede homemaking to women, widowed men also often find their meals irregular and their homes—if not their entire lives—in disarray.

We do not know yet about same-sex marriage. It could be that married homosexual men each gain seven years of life, and homosexual women each gain two years, just like heterosexual men and women. But it is also possible that married homosexual men gain two years, while homosexual women gain seven. If this were the case, it would mean that it is not marriage per se that is salubrious but, rather, marriage to a woman.

These differences in men and women highlight the fact that whom we are connected to may be just as important as whether or not we are connected. Two people might have different numbers of friends, or they might have the same number of friends, but one person might have educated friends and the other uneducated friends. This difference in the nature, and not just the number, of the social contacts itself is often significant.

For example, both the age and race of your spouse can have an effect on your health. Marriage to a younger woman is good for a man whereas marriage to a younger man is not good for a woman. A variety of investigators have shown that the bigger the age difference (up to certain limits) between an older husband and a younger wife, the better for both parties when it comes to the health benefits of marriage.[42] Some have interpreted this finding to be consistent with the "trading sex for cash" caricature: if marriage works to provide health by improving women's economic well-being and men's social well-being, then, on average, older men and younger women are better able to provide these benefits to each other.

What we are describing here are average effects, of course, and many people have different experiences. It is likely that in couples where the wife is the principal breadwinner and leaves behind substantial assets, and where the husband's role is to provide social connection, the death of a wife may not be as harmful to the husband's health. In fact, in more socially egalitarian societies, the widower effect may be more similar in men and women.[43] This is what we would expect if the gender roles of "breadwinner" and "social connector" create the relative health advantages for husbands and wives.

Men and women may also differ in their ability to offer and receive the benefits of marriage, which raises an important question: Do men benefit more from marriage than women because men stand to gain more from marital connection or because women have more to offer? Or both? One way that we approached these broader questions, in conjunction with sociologist Felix Elwert, was to study racial variation in the widower effect. We found that white couples suffer a widowhood effect, but black couples do not. There are a variety of possible explanations for this result, but the most compelling seems to be that the health benefits of marriage endure after the death of a spouse in black couples but not in white couples. But, if you are a white man, why do you fare worse than a black man during bereavement? Is it because you are white, or because you were married to a white woman?

We developed a large sample of interracial couples to sort this out and found that men married to black women did not experience a widowhood effect, whereas men married to white women did, regardless of the man's own race.[44] But how could a wife's race affect her husband's mortality during widowhood? Clearly, any effect on the husband's health cannot relate to the wife's ongoing efforts. She is dead, after all. Rather, the effect must be caused either by aspects of the dissolved marriage that vary between racial groups or by the characteristic circumstances of the state of widowhood that vary between racial groups. For example, maybe the families of black wives are more supportive during the bereavement of husbands, on average, than are the families of white wives.

This may in turn relate to the greater rejection of intermarried couples by white relatives than by black relatives. Since wives are typically responsible for maintaining kinship networks, black men married to white women may be more likely to suffer isolation, disconnection from their social network, and lack of support from their in-laws upon the death of their spouse than are black men married to black women or white men married to black women. Hence, the difference in marriage benefits between women and men is very likely to be a consequence of the greater ability of women to keep their spouses connected.

Love, Sex, and Multiplexity

Social networks function in large part by giving us access to what flows within them. We know for example that marriage to an educated, rich, or healthy spouse is better for our health than marriage to a person lacking these qualities. But this is not merely because of our spouse's identity, it is because of what they actually give. Healthy or educated or rich partners are better able to provide useful information, social support, and material goods. And the flow of love and affection between spouses is also critically important. One study of 1,049 couples followed for eight years found that a bad marriage accelerates the normal decline in health as people age. This is partly due to negative interactions with your spouse putting stress on your cardiovascular and immune systems, a kind of wear and tear that accumulates over time. As a result, the death of a spouse who does not love and care for you, or whom you do not love and care for, does not harm your health as much as the death of an intimate spouse.[45] It is little wonder, then, that we spend so much time searching for partners. The qualities of the people to whom we are connected will have a big effect on every aspect of our lives.

You may use lots of different kinds of social networks to help you in your search, whether they are networks of coworkers, Facebook friends, family members, or neighbors. The tendency to have several kinds of relationships (and sometimes many kinds of relationships with the same person) is called *multiplexity*. Our sexual network is in fact a subset of the larger social network within which we search for partners. In a sense, the latter is a potential network and the former is a realized network (like a network of contacts in a Rolodex, only some of whom become business partners).

If we live in multiplex networks, how we perceive them and how scientists draw them depends on what types of relationships we are focused on, as shown in plate 2. There are many layers, and your position in each layer will determine how connected you are. For example, you might have many friends but few sexual partners. This means you will be more central in the friendship network than you would be in the sexual network, even though both of these are a part of your complete social network. As a consequence, you are more likely to receive things that flow via friendship (like gossip) than things that flow via sexual relations (like sexually transmitted diseases). And some people will have multiple relationships to the same person, like the circled pair in plate 2.

We could use sexual interactions to trace paths through the network that would otherwise seem quite preposterous, or that might not be apparent, say, if we were instead tracing out paths involving business relationships. This observation even prompted American writer Truman Capote to develop a parlor game, which he described as follows: "It's called IDC, which stands for International Daisy Chain. You make a chain of names, each one connected by the fact that he or she has had an affair with the person previously mentioned; the point is to go as far and as incongruously as possible. For example, this one is from Peggy Guggenheim to King Farouk. Peggy Guggenheim to Lawrence Vail to Jeanne Connolly to Cyril Connolly to Dorothy Walworth to King Farouk. See how it works?"[46]

An important property of multiplex networks is that they overlap. We might be friends with our spouse, lovers with a coworker, or acquaintances with a neighbor. And when we seek out sexual partners, we typically draw on other kinds of networks. We do not form random ties with simply any other human. We do not choose our dates by throwing darts at the phone book. We get to know our neighbors, coworkers, schoolmates, and others to whom we are introduced, or, less often, meet serendipitously in a manner typically governed by other social constraints.

Thus, we might be able to learn a lot about social networks in general by looking at sexual networks in particular. And these networks are especially important because having sex with someone is clearly a very deliberate and detectable type of social tie. It is the analogue in social-network studies to what death is in medicine: an unambiguous end point. If we want to know who is connected to whom in a network, and we ask you who your friends are or whom you trust, these questions are much more open to vagaries of interpretation than asking whom you have had intercourse with. But by asking that question, it becomes possible to map out social networks in a well-defined way. And by probing how people find others to have sex with, and the other myriad ways that social networks affect our sex lives, we can understand more about human experience and social interaction than just sex. In the next chapter, we discuss how researchers have used sexual networks to study the spread of disease, and how this long-standing work set the stage to completely change the way we think about our health.

CHAPTER 4

This Hurts Me As Much
As It Hurts You

R ockdale County, Georgia, is a quiet upper-middle-class suburb about twenty miles outside of Atlanta. The schools there are some of the best in the state, and the annual fair draws large crowds to its church-choir nights and beauty pageants. Some natives of Rockland County have even become quite famous, like actresses Dakota Fanning and Holly Hunter. According to the county's website, this is a "'family-friendly' community that is appealing to parents who want a safe, wholesome, and progressive environment in which to raise their children."[1] In short, this is probably not the place you imagine when you think of a teenage syphilis epidemic.

But in 1996, very young teenagers started arriving at Rockdale County health facilities infected with syphilis and other sexually transmitted diseases (STDs). When officials began hearing sordid tales from teenagers in middle school and high school who participated in group sex, it became obvious that something strange was afoot. As Georgia's director of public health, Kathleen Toomey, told the *Atlanta Journal-Constitution,* "What lit this up for us [was that] syphilis had occurred in a community where you never see syphilis.... It allowed us to be aware of the high-risk behaviors of these teens in Rockdale."[2] Syphilis is an extremely rare disease in the children of the upper-middle class, yet there were seventeen cases of this one sexual pathogen alone, and many more cases of other STDs.

Some of the students of Rockdale County had accumulated dozens of partners. The epidemic, when it was discovered, made a big impression on the adults: "By the time the investigation concluded, seasoned public health investigators trained not to be judgmental would be startled by what they found. There were fourteen-year-olds who had had up to fifty sex partners, sixth-graders competing for the sexual attention of high school students, girls in sexual scenes with three boys at once. In one case,... a girl at a party with thirty to forty teens volunteered to have sex with all the boys there — and did. 'My heart dropped,' [Peggy] Cooper [a counselor at one of the middle schools involved] said. 'I felt nauseated. I wanted to cry.'"[3]

When this situation came to light, people began questioning what made the children in this wealthy community behave this way. It appeared that many of the young people suffered from not having much structure, supervision, or anything else to do. But really, the STDs were a reflection of a different network process: the spread of a norm among the teenagers that sex — and sex of a particular kind, involving multiple partners — was acceptable. The real epidemic, the one at the root of the STD epidemic, was an epidemic of attitudes. Syphilis was not the problem; it was a symptom of the problem.

The schism between the parents and the reality of their children's sexual activity was clear in the reactions of parents and other adults to the outbreak. As Toomey told the *Washington Post,* parents were in "tremendous denial" that the local youth were sexually active at all.[4] One nurse told the *Atlanta Journal-Constitution* that they "found a lot of lack of communication between children and parents.... Some didn't know their kids were sexually active. Some, even when presented with evidence, refused to believe it. One woman cussed me out and said she knew her child was a virgin, until I said no, her child was pregnant."[5]

A formal investigation involving interviews with ninety-nine of the teenagers, including ten who had been diagnosed with syphilis, reconstructed the network of people connected to one another sexually.[6] It found that those with syphilis were extensively connected in the middle of the web. Over time, the network expanded to include more people who were recruited into the sexual practices of the group, making it more likely that syphilis would be transferred to other kids. At the core of the sexual network was a collection of young white girls, most of whom were not yet sixteen. They would participate in group sex with various clusters of boys, connecting separate groups that might not have otherwise been linked. A year later, however, the network had splintered into many smaller networks, in part because of community efforts to address the problem. So while most of the kids remained sexually active, STDs were less likely to spread because the population was no longer interconnected. The epidemic stopped because the network changed.

Your Ex-Lover's Lover's Ex-Lover

A spate of research over the past decade has focused on the structure of social networks and their role in determining the spread of STDs. Because these studies involve the examination of traceable and easily detectable germs, and because sex is ipso facto proof of a connection between two people, such studies provide a useful vehicle for examining how pairs of people might come together to form more complex network structures and how this process in turn affects the social experience of individuals and the spread of things other than germs. Studies of STDs demonstrate the emergent quality of networks, that is, how phenomena of interest must be understood by studying the whole group rather than by studying individuals or even pairs of individuals. A person's risk of illness depends not merely on his own behavior and actions but on the behavior and actions of others, some of whom may be quite distant in the network.

Drawing from the Add Health data, sociologists James Moody and Katherine Stovel teamed up with Peter Bearman to map the complete sexual network of a midsized, predominantly white midwestern high school using information on reported romantic partnerships over an eighteen-month period. This high school, to which they gave the pseudonym "Jefferson High," was in a community that was superficially similar to Rockdale County. Moody and his colleagues found that a surprisingly sizable 52 percent of all romantically involved students were embedded in one very large network component that looked like "a long chain of interconnections that stretches across a population, like rural phone wires running from a long trunk line to individual houses."[7] This ring-shaped, hub-and-

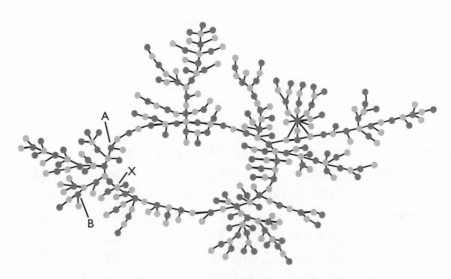

Network of 288 students involved in romantic relationships at "Jefferson High School" from the National Longitudinal Study of Adolescent Health. Gray nodes indicate girls, and black nodes boys. Nodes A, B, and X are discussed in the text.

spoke network composed of 288 students was especially notable for its lack of redundant ties, meaning that most students were connected to the superstructure by just one pathway, as shown in the illustration. There was not a lot of transitivity in the network.

Moody and his colleagues uncovered two particular rules regarding the social—in this case, specifically sexual—interaction in high schools that have a big effect on the structure of this network. First, there is a convention that people have partners that they resemble (our by-now handy homophily principle, here with regard to grade, race, and so on). Second, there seems to be a rule about sex in high school that says, Don't date your old partner's current partner's old partner.

We are pretty sure you had to read that rule a couple of times to make sense of it. We are also pretty sure you did not write it in capital letters at the top of your little black book when you were in high school. Nonetheless, if you consider all the partners you have had, we think you will be hard-pressed to discover a time where you violated this rule. One easy way to check is to ask yourself, "Have I ever swapped partners with my best friend?" Chances are, you have not.

This no-partner-swapping rule is an example of how social processes can determine overall network structure in a way that individuals themselves may not be able to appreciate or influence. That is, people obey a seemingly simple rule (consciously or not) and then wind up being embedded in a network with a particular structure. They cannot meaningfully influence the overall shape of the network, even though it certainly influences them by affecting who they will have sex with and whether or not they will be at risk for acquiring an STD.

One very interesting implication of the rule is that it seems to contradict the patterns we discussed in the previous chapter. Many people date a person who is originally three degrees removed from them, but here we have a rule against getting involved with the friend of a friend of a friend. What is the difference? As it turns out, the gender and specific sequence of sexual relations among these social contacts matters a lot. Most close friendships are between people of the same gender, so the normal course of events is for a heterosexual couple to introduce two of their friends (the man's male friend is introduced to the woman's female friend), who can then have a relationship if they choose. However, if a woman leaves her boyfriend to date her best friend's boyfriend, the jilted boyfriend and the (now probably ex-) best friend are not likely to form a new couple.

Aside from the swinger culture especially prominent in the 1970s, where people engaged in wife swapping or husband swapping, these four-way sexual relationships have always been rare in the United States. This might be because of what Moody and colleagues call the "seconds" problem. The two jilted lovers have both come in second in a competition to someone in the other couple. No one is interested in the battle for the bronze medal at the Olympics, and nobody wants to hook up with their ex-lover's lover's ex-lover.

Spreading Germs

Now that we've thought a bit about the structure of the network, let's think about how that structure affects flow. Imagine person X in the network of romantic partners shown in the illustration on page 98 gets an STD. Imagine that you are person A. You are five partners away from person X and have no real way to know what is happening in her life—who she is sleeping with, what she is thinking, whether she insists her partner use a condom, or what sexual practices she engages in. Yet, you are indirectly connected to this person, and the fact that she has acquired an STD has implications for your life that you will soon appreciate. The germ can spread from person to person and—in five hops—reach you.

Now, however, imagine that a tie two degrees away from you is cut, either because the connection between the two people dissolves (e.g., they no longer have sex with each other) or because they start using condoms (so contagion is interrupted even if a sexual relationship continues). Are you sure to avoid getting the STD? No, not really. Because this is a ring network, and you are on the ring. The structure (which you cannot appreciate unless you have the kind of bird's-eye view that we have here) is such that the STD can go around the other path of the ring and still reach you. Admittedly, it would take many more hops. But you are not immune from what is happening to others in your network, in some cases even if they stop having sex or use protection.

Now imagine that you are person B. Like person A, you are connected to three sexual partners. You are also five partners away from person X. But now, if a tie two people away from you is cut or interrupted, you are isolated from the epidemic. Your position in the network is actually quite different from person A's, but you do not have the perspective to see that. As far as you know, you have merely had sex with three partners, just like person A. Without such a complete view of the network, there really is no way for you to acquire that perspective. You are at the mercy of the network you reside in, with only a certain degree of control over who you are connected to directly—but with no control over who you are connected to indirectly.

To see why network structure matters, let's compare the Jefferson High network with another one that involves a similar number of people. The next illustration represents a network of 410 mostly adolescent men and women who were part of an epidemic of STDs occurring over a two-year period in Colorado Springs.[8] Just like the Jefferson High network, the Colorado Springs network shows that people obey the no-partner-swapping rule. However, the larger patterns of contact are much more complex, so that cutting one tie is less likely to remove a person from contact with the rest of the network.

Network of 410 mostly adolescent men and women who were part of an epidemic of sexually transmitted diseases occurring over a two-year period in Colorado Springs. Node A is discussed in the text.

For example, notice that in both networks person A has had three sexual partners, but in Colorado Springs each of those three partners themselves had many more partners. This increases the chance that one of person A's partners' partners will contract the disease and then possibly spread it to person A. So here is a simple rule: the more paths that connect you to other people in your network, the more susceptible you are to what flows within it.

Most models of disease transmission assume the existence of highly sexually active "cores" of individuals who disseminate disease to less active individuals at the periphery, and they assume that these cores may sustain epidemics by functioning as reservoirs of infection. For example, network approaches have been used to understand the racial differences in rates of STDs in the United States.

Sociologist Ed Laumann and his colleagues proposed that STD rates were higher among blacks than whites because of differences in the two groups' sexual network patterns.[9] A peripheral black person (where *peripheral* is defined as having only one sexual partner in the past year) is five times more likely to choose a partner in the *core* (defined as having four or more partners in the past year) than is a peripheral white person. No one has yet discovered why this is the case, but the result is that STDs would be more likely to be contained within the white core, whereas they are more likely to spill out into the black periphery.

In other words, whites with many partners tend to have sex with other whites with many partners, and whites with few partners tend to have sex with whites with few partners. This keeps STDs in the core of active white partners. On the other hand, blacks with many partners have sex with other blacks with many and few partners. Hence, STDs spread more widely through the black population.

A 2001 study used individuals' reports about their number of sexual partners to make inferences about the "Swedish Sex Network" (which sounds more like an X-rated movie than an academic investigation). It similarly suggested that the network had highly active cores.[10] Further, it concluded that safe-sex campaigns would be most effective if messages were directed at high-activity members (the cores, or hubs, of the networks) rather than targeted equally to all members of a community.

A network perspective also helps us get away from thinking that the main risk factors for STDs are individual attributes, such as race. In fact, a more effective approach to understanding risk is to focus on the architecture of a person's social network, namely their structural position rather than their socioeconomic position. We should not assume that money, education, or skin color cause people to engage in more or fewer risky behaviors. Studies of social networks are showing that people are placed at risk not so much because of who they are but because of who they know—that is, where they are in the network and what is going on around them. This structural perspective sheds new light on many social processes.

Different Network, Different Prescription

The syphilis epidemic in suburban Rockland County demonstrates that specific information about social-network structure and behavior—beyond the simplistic core versus periphery dichotomy or the sexually active versus sexually inactive dichotomy—is highly relevant to the spread of STDs. It shows that if the only link between two groups is removed, the transmission of infection between those groups is effectively halted as the network breaks into disjointed components. Some sexual networks (such as the one at Jefferson High) are highly vulnerable to the removal of a few ties or changes in individual behavior. Under these circumstances, the best prevention strategy is thus a broad-based, "broadcast" STD control program that targets the entire population rather than specific activity groups. Any one person's change in behavior can break the chain.

However, not all networks have the same shape, and therefore different strategies might work better for different groups. In studying the HIV/AIDS epidemic in sub-Saharan Africa, another group of researchers collected information on recent sexual partners of the residents of seven villages located on an island in Lake Malawi.[11] They found that, contrary to expectations, residents reported relatively few partners. There was no distinction to be made between core and periphery, and pretty much everyone had similar levels of sexual activity. The Malawi sex network did not have any high-activity hubs, that is, individuals or groups capable of sustaining the HIV/AIDS epidemic by having many sexual partners.

Despite these findings, however, on mapping the sexual network, the researchers discovered that a striking 65 percent of the population aged eighteen to thirty-five formed one large interconnected component, similar to the Colorado Springs network. Unlike Jefferson High or the Rockland County school networks, this network structure was strikingly robust against the removal of individual ties or nodes since there were numerous redundant paths (i.e., instances in which people directly or indirectly shared more than one sexual partner).

These findings call into question many assumptions about STD transmission in sub-Saharan Africa and elsewhere. The current epidemic is not being driven by a high-activity core made up of sex workers and their patrons, or by other high-activity individuals transmitting disease to a low-activity periphery made up of individuals with one or few partners. Simple enumeration of the numbers of partners per person could not yield these insights without actually mapping the networks.

In short, when trying to understand the spread of STDs, how and even whether the disease spreads depends on the larger patterns of contact in the overall network. Without information on individuals' partners' partners and their interconnections to other individuals in the population, we cannot determine whether a person is at high or low risk of contracting an STD. In fact, the situation is even more complex since ideally, in addition to the structure of the network, the way the ties and the overall network structure changes across time should also be taken into account.[12] Fortunately, scientists and doctors are becoming more serious about collecting network data, and they are developing techniques to visualize and analyze networks. This will help tremendously in the fight against HIV and other STDs. And it will allow us to study the spread of other health phenomena through social networks that are much less orthodox.

Your Friends' Friends Can Make You Fat

Germs are not the only things that spread from person to person. Behaviors also spread, and many of these behaviors have big effects on your health. For example, peers influence young people's eating behaviors, particularly weight-control behaviors in adolescent girls. And strangers can have an effect too: people randomly assigned to be seated near strangers who eat a lot wind up doing the same, and the effect can be so subconscious that it has been called "mindless eating."[13] It seems that we just can't help imitating others.

It turns out that we do not only imitate the people sitting next to us in a classroom or a dining room. We also imitate others who are much farther away. Similar to spreading germs, health-related phenomena can spread from person to person and from person to person to person, and beyond.

Our first effort to understand how this might work looked at obesity. We were drawn to the topic by the widespread claim that there is an obesity "epidemic" in the United States. This turn of phrase conjures up images of a plague out of control, and in fact, the word *epidemic* has two meanings. First, it means that there is a higher-than-usual prevalence of a condition. Second, it connotes contagion, suggesting that something is spreading rapidly.

It is readily apparent that the prevalence of obesity is rising. A standard measure of obesity is body mass index (BMI is weight in kilograms divided by height in meters squared). A normal BMI is considered to be in the range from 20 to 24, overweight is a BMI of 25 to 29, and obese is a BMI of 30 or more. From 1990 to 2000, the percentage of obese people in the United States increased from 21 percent to 33 percent, and fully 66 percent of Americans are now overweight or obese. What is not obvious is whether obesity can be seen as an epidemic in the second sense of the word. Is the epidemic more than metaphoric? Does obesity spread from person to person? And if so, how?

In order to study this question, we needed a special kind of data. These data were very difficult to come by because we needed to know about whole groups of people and also about their interconnections. We needed to develop data that contained precise information about people's positions in a large-scale network and about the architecture of their ties—who they knew, and who those people knew, and who those people knew, and so forth. We also needed to know people's weight and height and a lot of other information about them. And we needed this data across time with repeated observations on all the people in the network. No data set with all these attributes existed at the time we became interested in the obesity problem.

Undeterred, we decided to start with an epidemiological study known as the Framingham Heart Study that has been ongoing in Framingham, Massachusetts, just west of Boston, since 1948. Physicians have learned much of what is known about the determinants of cardiovascular disease from this famous study. When it was initiated, about two-thirds of all adult residents of Framingham signed up to be examined by doctors every two years and, amazingly, those who are still alive continue to be examined. While all of the people in the study originally resided in Framingham, many have since moved throughout Massachusetts and the United States. Their children and grandchildren also signed up to participate in follow-up studies starting in 1971 and 2001, respectively, and they too continue to be followed at regular intervals.

Quite by chance, we discovered that the Framingham Heart Study kept meticulous, handwritten records so that staff can reach participants every two to four years to remind them to come back for examinations. We could not believe our luck because these records— which had not previously been used for research purposes—included detailed information about the friends, relatives, coworkers, and neighbors of each participant. And since Framingham is a close community, a lot of the people who were relatives or friends or coworkers or neighbors of the participants were also, coincidentally, participants in the study. We were thus able to use these records to painstakingly reconstruct the social networks of all the subjects. Ultimately, we were able to map more than fifty thousand ties (not counting connections to neighbors) by focusing on a key group of 5,124 people within a larger network composed of a total of 12,067 people. We could also study how the ties changed from 1971 to the present and link the new social-network data to preexisting information about people's weight, height, and other important attributes.

To better understand these complex data, our first step was to map the network to see if we could visually discern any clusters of heavy and thin individuals, as shown in plate 3.[14] While clustering of obese and nonobese individuals is apparent in this graph, the pattern is extremely complicated. Consequently, we used special mathematical techniques to confirm that there was in fact substantial clustering of obese and nonobese individuals and that the clustering was not merely due to chance. In a surprising regularity that, as we have discovered, appears in many network phenomena, the clustering obeyed our Three Degrees of Influence Rule: the average obese person was more likely to have friends, friends of friends, and friends of friends of friends who were obese than would be expected due to chance alone. The average nonobese person was, similarly, more likely to have nonobese contacts up to three degrees of separation. Beyond three degrees, the clustering stopped.

In fact, people seem to occupy niches within the network where weight gain or weight loss becomes a kind of local standard. These niches might typically involve one hundred to two hundred interconnected individuals. This finding illustrates a more general property of large social networks: they have communities within them, and these communities can be defined not only by their interconnections but also by ideas and behaviors that their members come to share. These ideas and behaviors arise and are sustained in adjoining individuals and depend somewhat on the particular pattern of ties within the region of the network a person inhabits.

Our next challenge was to show that the clusters of obese and nonobese people in the social network did not result solely because people of similar weight tended to hang out together (homophily) or because people shared common exposures to forces that caused them to gain weight simultaneously (confounding)—the familiar issues confronted in the studies of the widower effect and other interpersonal effects. We wanted to see if there was a causal effect, meaning that one person could actually cause weight gain in others, in a kind of social contagion. One way we addressed the impact of homophily was straightforward: we simply included information in the analysis about the kinds of friends people chose, thereby taking the tendency of people to befriend similar people into account. But to deal with confounding required a different approach.

Suppose that Nicholas and James are friends. We ask James who his best friend is, and he says, "Nicholas." But then we ask Nicholas the same question, and he names someone else. This means that even though Nicholas and James are friends, James is probably more influenced by Nicholas than Nicholas is by James. And if Nicholas and James both name each other (they are mutual best friends), then they probably have a closer friendship than they would if just one of them named the other. So we should expect the strongest influence between mutual best friends.

Now suppose that confounding is the only source of similarity in weight between friends. If Nicholas and James start spending time in a new fast-food restaurant that just opened near them, they may both gain weight (a confounding effect), causing it to look like one of them is affecting the other. But they will gain weight regardless of who named whom as a friend. That means mutual friends, the friends we name, and the friends who name us will all appear to have equal influence. If, on the other hand, there are differences in the magnitude of the effects, then it suggests that confounding is not the only source of the similarity. A hamburger joint that is making both people gain weight does not care who named whom as a friend.

Variation by nature of the friendship tie is exactly what we found. If a mutual friend becomes obese, it nearly triples a person's risk of becoming obese. Furthermore, mutual friends are twice as influential as the friends people name who do not name them back. And finally, people are not influenced at all by others who name them as friends if they do not name them back. In other words, if Nicholas does not consider James his friend, then James has no effect on Nicholas, even if James considers Nicholas his friend.

In addition to friends, we found that weight gain could spread through a variety of social ties from person to person, but they had to be close relationships. Spouses and siblings influence each other. Coworkers influence each other too, as long as they work in a small company where everyone knows one another. These effects can also spread between people who are not close as long as the effects travel through a series of close relationships in the network. You may not know him personally, but your friend's husband's coworker can make you fat. And your sister's friend's boyfriend can make you thin.

The final step in our study was to make a series of video animations that tracked the evolution of weight gain and social-network connections across thirty-two years. When we began this work, we thought we would be able to literally see the epidemic unfolding. We had thought we would see one person gain weight and then watch a wave of obesity spreading out from the affected person: first to those one degree away, then two degrees away, then three degrees away, across time and across social space. The image in our heads was based on the physics experiment many people are familiar with: a pebble is dropped in a still pool of water, and a concentric circle of waves moves away from it. If the waves hit a perimeter, they bounce back, and under proper circumstances, they can reinforce each other; peaks and valleys of "standing waves" form, like swells observed off-shore that seem not to move. Similarly, we expected to see concentric regions of the social network with peaks and valleys of obese and nonobese people.

Yet when we looked at our video images, the picture was much more complicated. There seemed to be chaotic weight gain all over the place. And we then realized that the proper analogy was not a single pebble dropped in a pool, but rather a whole handful of rocks thrown in over a wide area, creating a choppy surface, obscuring the impact of a single pebble and its waves. Sure, obesity can spread, but it is not spreading from just one spot, and social contacts are not the only stimulus for weight gain. People take up eating, stop exercising, get divorced, lose a loved one, stop smoking, or start drinking, and each one of these changes can form the epicenter of another tiny obesity epidemic, like the thousands of overlapping earthquakes that shake our tectonic plates every year. The movie was telling us something very important: the obesity epidemic does not have a patient zero; it is not a unicentric epidemic but a multicentric epidemic.

When our findings that obesity is contagious were published in 2007, they elicited strong reactions. We received hundreds of e-mails and noted hundreds of blog posts about this work. Some people exclaimed indignantly, "Well, *of course* obesity can spread like a fad." After all, body types come in and out of fashion, don't they? One year the frail waif is in fashion, and the next year it's the buxom Brazilian supermodel. Any casual observer of military history can see the huge change in male body type between World War II and the Iraq War just by looking at photographs of soldiers awaiting deployment. "And, by the way," our critics noted, "what a colossal waste of money it is for social scientists to prove the obvious." But others had quite a different response, equally outraged, to the suggestion that something so personal, so individual, so *clinical*, as weight gain could remotely be subject to the vagaries of popular tastes. "Weight can't *possibly* be contagious!" they said, "Everyone knows that weight gain is a function of genes and hormone levels and all kinds of choices and opportunities that people face. There must be something wrong with your research. And, by the way, what a waste of money it was."

But we now know that obesity is contagious. Since the publication of our study, we and three other independent teams have identified obesity contagion in other populations.[15] And there is something both commonsensical and novel about this observation. But how is obesity contagious? What other health phenomena might spread in this way? And what are the implications of knowing that a key feature of our health depends on a key feature of the health of others near and far in our social networks?

Changing What We Do, or Changing What We Think?

Longshoreman and social critic Eric Hoffer once opined, "When people are free to do as they please, they usually imitate each other." And imitation is one way obesity might spread from person to person. If you start a running program, then your friend might copy you and start running; or you might invite your friend to go running with you. Similarly, you might start eating fattening foods, and your friend might follow suit; or you might take your friend to restaurants where you eat fattening foods together.

Behavioral imitation can be either conscious or subconscious. In chapter 2 we noted that when we see someone eat or run, our mirror neurons fire in the same part of the brain that would be activated if we ourselves were eating or running. It is as if our brains practice doing something that we have merely been watching. And this practice in turn makes it easier for us to exhibit the same behavior in the future. There are still other physiological processes underlying imitation, such as those pertaining to contagious yawning or laughter.[16] Imitation, in other words, can be cognitive (something we intentionally think about) and physiological (a natural biological process). It is deeply rooted in our biological capacity for empathy and even morality, and it is connected to our origins as a social species, as we will discuss in chapter 7.

But imitation is not the only way that obesity can spread. Human beings also share their ideas with one another, and these ideas can affect how much we eat and also how much we exercise. For example, we might look at those around us and see that they are gaining weight, and this might change our ideas about acceptable body size. When many people start gaining weight, it can reset our expectations about what it actually means to be overweight. What spreads from person to person is what social scientists call a *norm,* which is a shared expectation about what is appropriate.[17] Just as the Rockdale County teenagers adjusted their sexual norms (to the dismay of the adults), people's ideas about what is considered fat have changed rapidly. And niches within a social network can arise. In these niches, people can reinforce particular norms so that directly and indirectly connected people share an idea about something without realizing that they are being influenced by one another.

These two mechanisms—behavioral imitation and norms—can be found in some of the examples we gave in the previous chapter for how marriage affects health. But it can be difficult to tell them apart. When a man gives up his motorcycle after getting hitched, is he copying his wife's behavior (she doesn't have a motorcycle) or adopting a new norm (the infernal things are unsafe)? Moreover, having accepted the idea that weight gain is normative, a person might go on to exhibit either the same or entirely different weight-gain behaviors from those around him. People around you might eat poorly and consequently gain weight, but you might wind up exercising less rather than copying the bad eating habits. While the norm that spreads is the same in this scenario (weight gain is OK), the behavior is not. Thus, there can be a concordance of norms even if there is not a concordance of behaviors. The spread of obesity is not simply a matter of monkey see, monkey do.

In the case of obesity, there is evidence that normative influences are at work. First, the spread of obesity in social networks still occurs even after accounting for the interpersonal spread of particular behaviors that contribute to obesity. That is, even if we do statistical analyses that take into account the fact that two people might copy each other's behaviors, we still find evidence that something more is going on, that weight gain and loss still spread.

Second, obesity can spread even between socially close people who are very far apart geographically. Remarkably, our evidence in the Framingham Heart Study suggested that social contacts a thousand miles from each other can influence each other's weight. Since we only rarely see our friends and family who live so far away, it is unlikely that the effect results from simple imitation. Suppose, for example, that you see your brother once a year at Thanksgiving, and he has gained a lot of weight. Copying his eating behavior on that one day will not affect your long-run weight status. But seeing him in his new, bigger physique can reset your expectations about acceptable body size. "Wow," you say, "Dimitri has put on some weight, but he is still Dimitri." When you go home, you can still carry this idea with you ("Dimitri looks fine"), and it influences your behaviors; you might eat more, and you also might exercise less.

Norms can spread even if they do not affect a person's behavior. Some people can be carriers of an idea without themselves exhibiting the behavior related to the idea. As a result, you might seem to affect your friend's friend without affecting your friend. Think of it this way: Amy has a friend Maria, who has a friend Heather, and Amy and Heather do not know each other. Heather stops exercising and gains weight. Since Maria likes Heather, this influences Maria's thinking about what it means to be overweight, and Maria comes to think that it is not so bad to be heavy. Maria does not change her own behavior. However, she might become more tolerant of people who eat a lot or who do not exercise much. So, when Amy stops her exercise regime (she used to go running every week with Maria), Maria is less likely to pressure Amy to continue. Given the shift in Maria's ideas about weight gain, even if Maria's own behaviors have not changed, this affects Amy. Hence, Heather's actions can affect Amy even if Maria's actions do not change.

How can people detect and imitate local network norms about the acceptability of weight gain when our society as a whole still appears to privilege thinness? Celebrities and models are thinner than ever, even if everyone else is gaining weight. This paradox illustrates the difference between ideology and norms. People see images of ideal body types in the media, but they are less influenced by such images—by this ideology—than they are by the actions and the appearance of the very real people to whom they are actually connected. As columnist Ellen Goodman put it: "Professional anorexics such as Kate Moss, Calista Flockhart, and Victoria Beckham may present an incredibly shrinking ideal. But in real life we measure ourselves against our friends. Inch for inch."[18] As we will see in chapter 6, the same sort of thing happens with political beliefs.

It is worth stressing that social-network effects are not the only explanation for the obesity epidemic. Over the past twenty years, there have been enormous changes that promote inactivity—such as labor-saving devices, sedentary entertainment, suburban design, and the general transition to a service economy. There have also been dramatic changes in food consumption, resulting from the decreased price of food, shifts in nutritional content and portion size, and increased marketing. Yet social networks also play an important role. As we have discussed, networks can magnify whatever they are seeded with, though other external factors are the initial drivers of the obesity epidemic. If something takes root in a networked population, whether a pathogen or a norm about body size, it can spread across social connections, striking ever larger numbers of people.

How Smoking and Drinking Are Like
Back Pain and Koro

Connection is just as important when it comes to health phenomena beyond obesity. People copy the substance-use, drinking, and smoking behaviors of people they know directly and, more remarkably, of others who are farther away in the social network. Just as understanding social networks helps us understand the sharp rise in obesity in our society, it can also help us understand the sharp decline in smoking, the relative persistence of drinking, and a wide variety of other activities that affect our health.

Over the past forty years, smoking among adults has decreased from 45 percent to 21 percent of the population. Whereas forty years ago, offices, restaurants, and even airplanes were thick with cigarette smoke (the rule against smoking in airplanes was hailed as a great advance in 1987), nowadays smokers huddle together in small groups outside buildings.

But people have not been quitting by themselves. Instead, they have been quitting together, in droves. We used our social-network data from the Framingham Heart Study to analyze the decline of smoking over the past four decades, and we found patterns that looked like the obesity epidemic in reverse.[19] When one person quits smoking, it has a ripple effect on his friends, his friends' friends, and his friends' friends' friends. As in the case of obesity, smoking behavior extends out to three degrees of separation, consistent with our Three Degrees of Influence Rule. But the group effects are even stronger for smoking than obesity. There is a kind of synchrony in time and space when it comes to smoking cessation that resembles the flocking of birds or schooling of fish. Whole interconnected groups of smokers, who may not even know one another, quit together at roughly the same time, as if a wave of opposition to smoking were spreading through the population. A smoker may have as much control over quitting as a bird has to stop a flock from flying in a particular direction. Decisions to quit smoking are not solely made by isolated individuals; rather, they reflect the choices made by groups of individuals connected to one another both directly and indirectly.

Anthropologists have a word for local customs: *culture.* But the culture we are discussing here is local in the sense of being confined to groups of interconnected people in one region or niche of the social network rather than in one geographic place or among one group defined by a shared religion, language, or ethnicity. And the culture within regions of the social network can change. Interconnected groups of individuals may find smoking unacceptable and quit in a coordinated fashion, mutually influencing one another but without necessarily knowing each other personally or explicitly coordinating their behavior. What flows through the network is a norm about whether smoking is acceptable, which results in a coordinated belief and coordinated action by people who are not directly connected. This is an important way that individuals combine to form a superorganism.

Smoking behavior reflects the workings of a superorganism in still other ways. First, people who persist in smoking find themselves progressively marginalized in the network, as shown in plate 4. In 1971, smoking had no bearing on social position: smokers were just as likely to be central to their local networks as nonsmokers, to have as many friends as nonsmokers, and to be located in the middle of large extended groups. However, as more and more people quit smoking over time, the smokers were forced to the periphery of their networks, just as they are now forced outdoors to smoke, even in the freezing cold. And it's not just that they became less popular; they also tended to be friends with people who were less popular, which helped to speed up the dramatic increase in their social isolation.

Second, although smokers and nonsmokers were well mixed in the early 1970s, over time they each formed their own cliques within the network, with progressively less interconnection by the early 2000s. As in the case of the polarization between Democrats and Republicans in the U.S. Congress (discussed in chapter 6), the separation between smokers and nonsmokers has increased over time, and the consequences extend far beyond the habit itself. When such deep divisions occur, they lead to the formation of identities within each group, which prevents further mixing and reinforces group behavior. More connections within groups (in what is known as a *concentrated network*) can reinforce a behavior in the groups, but more connections between groups (in what is known as an *integrated network*) can open up a group to new behaviors and to behavioral change—for better or for worse.

The spread of smoking cessation also illustrates the role of high-status individuals in the diffusion of innovations. In our Framingham data, education appears to amplify a person's ability to influence others: a person is more likely to quit smoking when a well-educated social contact quits. In addition, education increases a person's desire to innovate: a well-educated individual is more likely to imitate the smoking-cessation behavior of his peers than is a less-educated individual. Hence, ironically, in the case of smoking, the present wave of cessation mirrors in the obverse what happened sixty to one hundred years ago: when smoking first took root in our society, it did so among those with relatively high status. Ads from the 1930s and 1940s show smiling doctors enjoying and promoting tobacco.[20]

Just as the educational status of individuals in the network is related to the spread of smoking along paths through the network, gender is relevant to the spread of drinking. The Framingham social network reveals that drinking behavior clusters to three degrees of separation, as do smoking and obesity. But the flow of influence does not pass through everyone equally—instead, drinking appears to be greatly influenced by women. If a woman starts drinking heavily, both her male and her female friends are likely to follow suit. But when a man starts drinking more, he has much less effect on either his female friends or his male buddies down at the bar. It is not yet clear why this is so, but it suggests that women are the key to the spread throughout the network, and they may therefore be the key to successful interventions.

Drinking has been relatively steady in the United States, but the situation is not so stable in other countries. For example, in the United Kingdom, there is a new problem with public binge drinking. Increasing numbers of young men and women are rapidly consuming large amounts of alcohol (as many as ten drinks per sitting) and then engaging in public behaviors such as vomiting, collapsing, urinating, shouting, threatening, and fighting. Overall, 16 percent of eighteen- to twenty-four-years-olds in one British sample reported having engaged in binge drinking.[21] Among binge drinkers, 54 percent reported that all or almost all of their friends are binge drinkers, compared to 15 percent of nonbinge drinkers. Analysis of these patterns suggests that there is indeed an interpersonal clustering and transmission of the behavior.

While gender and education have an effect on the spread of health behaviors, the type of relationship also matters a great deal. Not all social ties are equal. For example, we find that friends affect each other more than spouses do in the spread of obesity. When we first noted this, we were puzzled because spouses often eat together, exercise together, and spend more time with each other than friends do. However, when we looked more deeply, we found that friends and siblings are much more susceptible to influence by peers of the same sex than by peers of the opposite sex. Thus, although spouses are typically friends, they are also typically of the opposite sex, and the two effects can cancel each other out.

Other health-related behaviors that might spread within social networks include the tendency to get health screenings, visit doctors, comply with doctors' recommendations, or even visit particular hospitals. One study found that Harvard students were 8.3 percent more likely to get a flu shot if an additional 10 percent of their friends got a flu shot.[22] Moreover, symptoms can spread from person to person due to biological and social mechanisms. In chapter 2, we saw that anxiety and happiness can spread, but so can headaches, itches, and fatigue.

Back pain is yet another example of a condition that can spread via social networks. A group of German investigators studied the possible transmission of back pain by exploiting a natural experiment provided by the reunification of Germany. Before the Berlin Wall fell, East Germany had much lower rates of back pain than West Germany, but within ten years of reunification, rates had converged to be the same, with East Germany emulating West Germany's higher rates. Exposure to new media messages among formerly insulated East Germans about how back pain was "frequent and unavoidable" and "a diagnostic and therapeutic enigma in need of careful medical attention" appeared to play a role. But these investigators also argued that back pain was a "communicable disease" and that a kind of "psychosocial decontamination" might be helpful to break the transmission.[23] Thinking of back pain in this way can shed light on another mystery: it can help explain why prevalence rates vary widely among industrialized countries. The rate of lower back pain among working-age people is 10 percent in the United States, 36 percent in the United Kingdom, 62 percent in Germany, 45 percent in Denmark, and 22 percent in Hong Kong.[24]

In some ways, this varying prevalence, and the culturally specific ways in which back pain is experienced, suggest that back pain can be seen as a *culture-bound syndrome*—a disease recognized in one society but not others, such that people can experience the disease only if they inhabit a particular social milieu. The classic example of a culture-bound syndrome is *Koro,* a condition that is seen in some Asian countries and that involves intense anxiety arising from the conviction among afflicted men that their penises are receding into their bodies and might disappear, and that they might die as a result. The treatment consists of asking trusted family members to hold the penis twenty-four hours a day for some number of days to prevent it from receding. To the eyes of outsiders, this condition has no biomedical or clear etiological basis. Yet it is very real to those who suffer from it. Indeed, there have even been epidemics of Koro, as documented in Malaysia and southern China (where it goes by the name of *suo yang*). To Malaysians—who probably have a low prevalence of back pain—the fact that many Americans have a biomedically difficult-to-diagnose condition that requires those affected to miss work and that usually has no objective physical signs may strike them as equally inexplicable.

A similar argument can be made that anorexia and bulimia are culture-bound syndromes. These conditions are much more prevalent in wealthy industrialized countries, and, even within these countries, they tend to strike white, middle-class, adolescent girls with much greater frequency than other groups. Their prevalence has been increasing since 1935; roughly 0.5 to 3.7 percent of American women have suffered from anorexia and 1.1 to 4.2 percent from bulimia (rates in men are about a tenth of these).[25] While these conditions are entirely real for their sufferers and their families, their origins are obscure. What triggers the eating behavior? In addition to being culturally specific, eating disorders resemble other culture-bound syndromes in that they can ripple through a social network in waves, reflecting the possibility of person-to-person transmission of (admittedly severe) weight-loss behavior. High-school girls may compete with one another to lose weight and college dormmates can copy one another's binge eating. In fact, these behaviors may affect a person's network location, and in one study of sororities, women who were binge eaters actually became more popular and moved to the center of the social network, just as nonsmokers did in our study.[26] As such, epidemics of eating disorders are an extreme example of the transmission of weight behaviors we documented in the Framingham Heart Study population.

Contagious Suicide

Suicide contagion is perhaps the most devastating illustration of the power of social networks. There are many causes of suicide, but the idea that people could kill themselves simply because others do seems to defy rational explanation. It certainly calls into question the whole notion that suicide is merely an individual act.

Suicide clusters have occurred throughout the world in communities of all types—rich and poor, big and small. Examples are even known from antiquity. While some clustering of suicides might be expected to occur due to chance, many of the clusters reflect contagious processes and are not due to chance, confounding factors, or homophily (among people who somehow have a prior inclination to kill themselves).[27] These clusters are different, in other words, from ones that are organized by charismatic cult leaders like Jim Jones, who led more than nine hundred of his followers in mass suicide in 1978 (a particularly powerful example of both confounding and homophily).

The classic investigation of suicide contagion was published by sociologist David Phillips in 1974.[28] He showed that during the period from 1947 to 1968, suicides increased nationally in the month after a front-page article appeared in the *New York Times* describing someone who had taken his own life. Phillips dubbed this "the Werther effect," after Johann Wolfgang von Goethe's novel *The Sorrows of Young Werther* published in 1774. The novel was read widely, and when some young men began committing suicide in a way that copied the protagonist, authorities in Italy, Germany, and Denmark banned the book.

There are two kinds of suicide cascades: those that work through media contagion, like *Young Werther* or the front page of the *Times* (these can involve either fictional or factual accounts), and those that work through direct contagion among people who are connected to a person who has killed himself.

Concerns about media contagion have been sufficiently serious that they have led the Centers for Disease Control to suggest alternative ways of publicizing the occurrence of suicide.[29] The CDC has even promulgated sample obituaries for journalists. Here is the type of news report the CDC feels has "high potential for promoting suicide contagion":

> Hundreds turned out Monday for the funeral of John Doe Jr., 15, who shot himself in the head late Friday with his father's hunting rifle. Town Moderator Brown, along with State Senator Smith and Selectmen's Chairman Miller, were among the many well-known persons who offered their condolences to the City High School sophomore's grieving parents, Mary and

John Doe Sr. Although no one could say for sure why Doe killed himself, his classmates, who did not want to be quoted, said Doe and his girlfriend, Jane, also a sophomore at the high school, had been having difficulty. Doe was also known to have been a zealous player of fantasy video games. School closed at noon Monday, and buses were on hand to transport students who wished to attend Doe's funeral. School officials said almost all the student body of twelve hundred attended. Flags in town were flown at half-staff in his honor. Members of the School Committee and the Board of Selectmen are planning to erect a memorial flagpole in front of the high school. Also, a group of Doe's friends intend to plant a memorial tree in City Park during a ceremony this coming Sunday at 2:00 p.m.

Doe was born in Otherville and moved to this town 10 years ago with his parents and sister, Ann. He was an avid member of the high school swim team last spring, and he enjoyed collecting comic books. He had been active in local youth organizations, although he had not attended meetings in several months.

And here is a suggested news report that the CDC feels has "low potential for promoting suicide contagion":

John Doe Jr., 15, of Maplewood Drive, died Friday from a self-inflicted gunshot wound. John, the son of Mary and John Doe Sr., was a sophomore at City High School.

John had lived in Anytown since moving here 10 years ago from Otherville, where he was born. His funeral was held Sunday. School counselors are available for any students who wish to talk about his death. In addition to his parents, John is survived by his sister, Ann.

Helpfully, the CDC notes that "the names of persons and places in these examples are fictitious and do not refer to an actual event."

Perhaps the CDC wants to avoid the possibility of inducing suicide even with fictitious reports. The point is that the second news report omits the personal and sympathy-inducing elements of the first. The CDC guidelines recommended that news reports not explain the method of suicide or mention how "wonderful" the deceased teenager was; they should also refrain from suggesting that the suicide helped solve the teenager's problems, for example, by getting even with Jane ("When contacted, Jane sobbed as she reported how much she missed John").

This works. When Vienna, Austria, finished its subway system in 1978, it was not long before people started using it for a purpose for which it was not intended: they flung themselves in front of the trains. Media reports were vivid, and suicide attempts (half of which were successful) numbered nearly forty per year. Viennese psychiatrists became concerned and started working with journalists. Changes in the reporting of suicides were implemented in 1987, and there was an immediate and enormous drop in suicide attempts to roughly six per year thereafter.[30]

Since Phillips's 1974 paper, the sophistication of suicide analysis has risen tremendously, and the geographic scale has narrowed to focus on localized outbreaks and those that occur by direct contagion. As in the case of MPI, the burden seems to fall especially heavily on schools and on small communities that are, as the saying goes, "tightly knit." Moreover, suicide contagion occurs almost exclusively among the young. Adults older than twenty-four show little, if any, excess likelihood of killing themselves if someone they know has done so or if they simply read about a suicide in the paper.[31] But teenagers, who are especially impressionable and susceptible to peer effects in so many domains of their lives, are another matter. The link between the subject's age and susceptibility is yet another illustration of how the attributes of the nodes on a network are crucial in determining the flow of the phenomenon at hand.

Here is how such outbreaks unfold. The average suicide rate in Manitoba, Canada, is 14.5 cases per 100,000, but in 1995, in a village of 1,500 people in the far north, the rate was 400 per 100,000. Six young people took their lives, mostly by hanging, in four months. A further nineteen attempted suicide. A sense of the epidemic, and how it spread through personal connections between the people in this small town, can be appreciated by the urgent report of one of the doctors who was flown in to help care for this community. This is his description of the events at the local health clinic during a three-day period that began two weeks after the last of the six successful suicides, when serious lingering effects were still observable:

A nineteen-year-old man presented to the health centre two weeks after the sixth suicide. The police were worried about him. "Three of my friends passed away, and I can't take it anymore." He had tried to hang himself in his bedroom two weeks earlier. His brother and a friend had discovered him and cut the rope. After a cousin died on the winter road the previous year, this man had tried to shoot himself and was prevented from doing so by his parents. He spent the night in jail, took to his room for a week, and "felt better." At the time of assessment, he acknowledged hearing the voices of two of the victims beckoning him to come with them. This occurred mostly when he was alone and was frightening. Vegetative symptoms of depression were not present. He requested an opportunity to "talk and get things out."

A thirteen-year-old boy was seen the same day because his father was worried about him. Collateral history revealed that the first suicide victim was the boy's cousin; he had discovered the second victim still hanging. His brother-in-law was the third victim. He denied suicidal ideation and had not attempted self-harm. The patient did not want to go to school anymore. He felt lonely, and dreams about the deceased frightened him. Playing hockey with a brother and cutting wood with his father were his favorite activities.

The following day, a fifteen-year-old girl was assessed. She was silent for the first fifty minutes before disclosing that two of the victims were her cousins. She acknowledged having heard her cousins' voices beckoning in the past but not for the last three weeks.

A twenty-three-year-old woman presented later that day. She had increased her alcohol consumption since the suicides. At one point she had written a suicide note, but the third victim, her uncle, "beat me to it and stole the show." She burned the note.... The second suicide victim was the niece of her boyfriend. The patient heard someone calling her name....

A fourteen-year-old girl, who was a friend to four of the victims and a cousin of a fifth, was brought in by her mother. She had had a dream that her cousin was smiling at her while he was hanging from a rope. One month earlier, she had been sent out for assessment after a hanging attempt. She had had previous suicide attempts.

A fourteen-year-old boy was seen next. He had tried to hang himself four months previously. All six victims were known to him. One was a cousin. Prior to his attempt, he had a dream about "a woman with long hair, a little bit spiky on top, with a black face and a long coat." He said, "Everyone here sees that woman at night." This boy also experienced someone beckoning but could not say who it was.

Later that night, a fourteen-year-old girl was brought in by the...constables....At 9:00 p.m. she had taken seven glyburide tablets [a medication for diabetes] and then told a friend. Both of the females who committed suicide were friends. One week earlier she had seen one of these girls in a dream telling her to kill herself.[32]

The foregoing record is overwhelmingly depressing just to read. One can readily imagine what it must have been like for the people in this village to be in the grip of the epidemic.

Another well-documented outbreak took place in a high school of 1,496 students in Pittsburgh, Pennsylvania. Two students killed themselves within four days, apparently prompted by the suicide of a twenty-one-year-old former schoolmate; and during an eighteen-day period that included these two cases, an additional seven students attempted suicide, while another twenty-three reported thinking about killing themselves.[33] It was possible to trace out transmission of suicide since the first high-school student was a friend of the ex-student and an acquaintance of the subsequent student who killed himself. Moreover, many of the kids who thought about killing themselves, or who tried to do so without success, had demonstrably close social ties to the kids who succeeded and to each other. Although many of the kids in this cluster had a prior history of depression, many did not. This brings up a key issue with respect to suicide cascades: does knowing someone who has taken her life merely encourage others who would eventually have attempted suicide to follow suit, or does knowing someone who has killed herself recruit new victims to the epidemic? This question is analogous to the one we considered in chapter 3 with respect to fertility cascades, and there we saw that a sibling having a baby did not just accelerate a person's having a baby, it actually increased the total number of babies a person had.

Direct contagion can work the same way for suicide that it does for obesity, that is, via a spreading of ideas rather than by shared behaviors. Suicide in one person may lower the threshold for others to follow suit by changing attitudes and norms. It may increase the sense that it is something desirable to do ("look how all these people are so sad at the death of that person"). A case of suicide may make a person feel that the usual normative pressure to refrain from killing oneself is partially suspended. Suicide in a familiar person may also provide information about how to do it. Of course, in some cases, it may even involve collaboration (as in documented cases of Internet suicide clubs in Japan, the United Kingdom, the United States, and many other developed nations, which are formed by two or more strangers for the purpose of killing themselves together or simultaneously).[34]

The most recent examinations of suicide cascades have merged network methods and very large data sets to further investigate and confirm direct contagion. A study of 13,465 adolescents in Add Health confirmed that having a friend who committed suicide increased the likelihood of suicidal ideation. Boys with a friend who had killed himself in the previous year were nearly three times as likely to think about killing themselves than they otherwise would be and nearly twice as likely to actually attempt it. Girls with a friend who had killed herself were roughly two-and-a-half times more likely to think about killing themselves and also nearly twice as likely to actually attempt it. But, with the Add Health data, it was possible to examine a host of other features related to a person's position in the social network. In addition to being stimulated by the suicide of a friend, other social-network structural risks of suicide include having fewer friends and being in a situation where your friends are not friends with one another (namely, having lower transitivity in your network). Adolescent girls (but not boys) whose friends are not friends with one another are subject to possibly conflicting norms about how to live their lives, and this may be stressful. It increases suicidal ideation more than twofold.[35] It's a sort of "If you two can't get along, I am going to kill myself!" effect.

Suicide contagion is not entirely unknown in adults. One study of 1.2 million people living in Stockholm during the 1990s found that men (but not women) who had coworkers who killed themselves were 3.5 times more likely to commit suicide than they otherwise would have been.[36] Interestingly, just as with obesity, which we found to be transmitted between coworkers only at relatively small firms, one person killing himself appeared to increase the risk of others doing so only in firms with fewer than one hundred employees. A person is more likely to have a real connection to the victim at a smaller firm than at a larger one.

There has actually been a smoldering but broad-based epidemic of suicide in the United States over the past few decades. A 1997 study found that 13 percent of American adolescents seriously considered suicide in the previous year, and 4 percent of adolescents actually attempted it.[37] Moreover, 20 percent of adolescents reported having a friend who had attempted suicide in the previous year. From 1950 to 1990, the rate of successful suicide for people fifteen to twenty-four years of age increased from 4.5 to 13.5 per 100,000.[38] Interestingly, there has also been an epidemic of fictional suicides over the same time frame. According to one analysis of movie plots drawn from an Internet movie database, IMDB.com, the percentage of total movies featuring a suicide rose from about 1 percent in the 1950s to over 8 percent in the 1990s.[39] Whether these two increases are connected, and which came first, is hard to know for sure. But it is clear that connections that can make us happy can also make us suicidal.

A New Foundation for Public Health

"You make me sick" is a colloquialism, but it reflects a reality. Our health depends on more than our own biology or even our own choices and actions. Our health also depends quite literally on the biology, choices, and actions of those around us.

This claim may strike some as anathema. Particularly in the United States, we are accustomed to seeing our destinies as largely in our own hands: we "pull ourselves up by our bootstraps" and believe that "anyone can strike it rich." We see our society as a meritocracy that rewards sound choices and creates opportunities for the well prepared. The radical individualist perspective is that we are masters of our destiny, and that by making changes in everything from what we eat to how we brush our teeth, we can improve our survival chances, our mental stability, or our reproductive prospects.

But the picture is much more complicated. Our unavoidable embeddedness in social networks means that events occurring in other people—whether we know them or not—can ripple through the network and affect us. A key factor in determining our health is the health of others. We are affected not only by the health and behavior of our partners and friends but also by the health and behavior of hundreds or thousands of people in our extended social network.

Most people know little about how the health of the public is protected. And what we do know, we think about in very self-oriented terms: the surgeon general's warning on the side of a cigarette pack or the nutrition labels on foods are targeted at individual users, not at the community as a whole. We do not ordinarily appreciate the ways in which one person choosing certain behaviors affects the health of others and why this provides a basis for public health.

Yet, we know that smoking-and alcohol-cessation programs and weight-loss interventions that provide peer support are more successful than those aimed at solitary individuals. Programs like Weight Watchers and Alcoholics Anonymous work precisely in this way: they cultivate the formation of social ties and group solidarity. Experiments confirm the beneficial impact of such interventions. One study randomly enrolled subjects in a weight-loss intervention in one of three ways: subjects could enroll alone, they could be assigned to a team of four people, or they could enroll as part of a team of four people that they formed (a method similar to that used for microloans to the poor, as will be discussed in chapter 5). Weight loss was 33 percent greater and also more durable when people were part of a group.[40]

Other experiments have also confirmed the interpersonal flow of health phenomena. For example, one study randomly assigned 357 people to receive a weight-loss intervention or not, but—unusually—it followed the 357 spouses of the subjects. Not only did the subjects in the weight-loss intervention lose weight, so did the spouses.[41] The primary mechanism for this was that the untreated spouse copied the eating behavior of the treated spouse, though a broader range of mechanisms was also likely.

A social-network perspective gives new credence to such group-level and family-level interventions, and it confirms that such interpersonal health phenomena might operate on a much larger scale. A network perspective demands a rethinking of the ways we as a society approach health and health care, and it suggests some novel approaches to public health.

Networks could possibly be manipulated in terms of the pattern of connections or the process of contagion so as to foster individual and collective health. If network ties could be discerned on a community-wide scale (for example, using some of the new telecommunications technologies and methods we will describe in chapter 8), we would be able to target influential individuals or those most at risk for being affected by interpersonal health processes. Moreover, if we knew people's ties on a large scale, we could design interventions to target groups of interconnected people.

As we have seen, people are more influenced by the people to whom they are directly tied than by imaginary connections to celebrities. Network science offers better ways to identify influential individuals by identifying centrally located hubs within the network.[42] In order to do this optimally, the entire network must first be drawn. For example, if we were trying to reduce smoking in a high school or workplace, the conventional approach might be to either broadcast a message to everyone or work with a small group that was felt to be especially at risk. In the latter case, these individuals might be identified because they were the poorest, say, or because they were known to be smokers already. But an alternative approach would be to identify the hubs in the social network (who might or might not be poor or smokers) and target them with smoking-cessation messages. Early results with such approaches have documented success in fostering better diets and safer sex.[43]

Yet, this approach shifts the focus of decades of public health work. It targets neither socioeconomic inequality nor socioeconomic or behavioral vulnerability per se, but rather structural inequality and structural vulnerability. People are placed at risk for bad or good health by virtue of their network position, and it is to this position that public health interventions might beneficially be oriented. In addition to focusing, for example, on whether people are poor or where they live, we might focus on who they know and what kinds of networks they inhabit.

Some recent work has clarified the specific circumstances whereby influential individuals are most apt to be able to exercise their influence. It turns out that influential people are not enough: the population must also contain influenceable people, and it may be that the speed of diffusion of an innovation is more dependent on the properties and number of the latter group than the former.[44] The key point, however, is that networks with particular features and topologies are more prone to cascades, that both types of people are required for cascades to take place, and that understanding the shape of the network is crucial to understanding both naturally occurring and artificially induced cascades.

Whether influential people can exercise influence at all may depend entirely on the precise structure of the network in which they find themselves, something over which they have limited control. As we have seen, some networks permit wide-reaching cascades, and others do not. If we light a tree on fire, whether this turns into a conflagration or a campfire depends a lot on what is going on around the tree: how close it is to other trees, how dry the terrain, how large or dense the forest. When the right conditions for a huge fire exist, any spark will set it off, but when they do not, no spark will suffice.

Computer models of the obesity epidemic confirm that targeting central individuals in the network to encourage healthy weight can be an effective strategy, whether these central people are overweight or not.[45] But these models suggest an even more unusual strategy:

at both the individual and the population levels, it is more effective for you to lose weight with friends of friends than with friends. The problem is this: If you attempt to lose weight with your friends, you might succeed, but this tiny cluster of you and your friends is surrounded by a large group of people exerting pressure to gain weight again. In all likelihood, both you and your friends will thus regain weight.

A good strategy to lose weight, therefore, might be to invite your friends to dinner and ask them to nominate their friends, and then invite those people to join a running club. If you were able to do this, you would also create a social force pressuring your friends to lose weight (since they would be surrounded), and you would create a buffer around you of people who are improving their health behavior.

Understanding networks can lead to still other innovative, non-obvious strategies. Randomly immunizing a population to prevent the spread of infection typically requires that 80 to 100 percent of the population be immunized. To prevent measles epidemics, 95 percent of the population must be immunized. A more efficient alternative is to target the hubs of the network, namely, those people at the center of the network or those with the most contacts. However, it is often not possible to discern network ties in advance in a population when trying to figure out how best to immunize it. A creative alternative is to immunize the acquaintances of randomly selected individuals.[46] This strategy allows us to exploit a property of networks even if we cannot see the whole structure. Acquaintances have more links and are more central to the network than are the randomly chosen people who named them. The reason is that people with many links are more likely to be nominated as acquaintances than are people with few. In fact, the same level of protection can be achieved by immunizing roughly 30 percent of the people identified by this method than would otherwise be obtained if we immunized 99 percent of the population at random! Similar ideas can be exploited for the opposite problem, namely, how best to conduct surveillance for a new behavior or a new pathogen (or bioterror attack): do we monitor people randomly or choose them according to their network position? A choice informed by network science could be seven hundred times more effective and efficient.[47]

Finally, network interventions increase the cost-effectiveness of interventions. For every dollar we spend on improving the health of an employee, we also improve the health of that employee's relatives, coworkers, friends, and even their friends of friends. This substantially increases the return on the investment. And, in the case of employers or insurers, this can be especially important since roughly two-thirds of workplace health costs are related to health problems in spouses and other dependents of workers. Targeting a worker and improving the health of the worker's family in the bargain is thus good business. As we will see in the next chapter, there are many ways networks can magnify economic benefits besides health care, and our understanding of economic behavior requires us to come to terms with the idea that no man or woman is an island. People are connected, and their health and well-being are connected.

CHAPTER 5

The Buck Starts Here

Not since 1866 had England seen such a crisis. In the summer of 2007, the worldwide real-estate bubble was bursting, the mortgage markets were grinding to a halt, and British banks were finding it harder and harder to raise funds in the money markets. Mortgage lenders were particularly hard-hit, and some were running out of options. On Wednesday, September 12, the Northern Rock bank shut its doors and asked the state-run Bank of England for money to cover its deposits. The news spread rapidly over the airwaves and by word of mouth. The government issued a statement saying customers should not be worried about their current accounts or mortgages, but to no avail. By Friday, September 14, when the bank's doors reopened, a real bank run was in progress for the first time in more than 140 years.

Long lines started forming outside Northern Rock branches throughout Britain as early as 6:00 a.m. Some customers came because everything they had was in a bank that appeared to be failing. "I have been saving for years, and I don't want to lose it," said

Jacqueline Porte, who had advanced just twenty-five feet toward the entrance of the Golders Green branch in three hours when she spoke to reporters.[1]

Others had less information about the bank's situation, but they came anyway because they saw the long lines on the television news or as they passed local branches. One customer who called herself Marilyn claimed she was reassured by the statements made by the government, but she could not resist the urge to join the run when she saw everyone else withdrawing their money: "I thought if I didn't come down here, I'd regret it."[2] A man in his fifties who preferred to remain anonymous said: "I'm an accountant, I should know better.... I shouldn't be here.... My head tells me it's all right, but my heart says otherwise."[3] Customer Anne Burke, fifty, waited with her ninety-year-old father in a line of 130 people outside the Brighton branch. "It's not that I disbelieve Northern Rock.... But everyone is worried, and I don't want to be the last one in the queue. If everyone else does it, it becomes the right thing to do."[4]

The run on Northern Rock also attracted people who did not have deposits at the bank. Tim Price, a portfolio manager, made a special trip to see the long lines of middle-class people waiting to withdraw their money. "It was a very British bank run," he said. "The queues were orderly, but the emotional impact will scar people for generations."[5] Others agreed with him. A mobile billboard advertising a suicide-prevention counseling service was parked in front of the Edinburgh branch. And not to be outdone, staff from other banks hovered outside several branches like vultures, with flyers that specifically targeted the fears of Northern Rock depositors, urging customers to switch accounts.

Meanwhile, Northern Rock management summoned extra staff and extended bank hours to deal with the continuing long lines and disgruntled customers. At one branch in Newcastle, customers burst into laughter when a staff member asked, "Does anyone want to pay money in?"[6] But at other locations things were becoming more tense. The Strathclyde police had to shut down one branch as a way to deal with "boisterous" customers. And in Cheltenham, police were called in to deal with a couple that barricaded a bank manager in her office, demanding that she let them withdraw one million pounds they had in their account.

These interventions did not stop the panic. The run continued for three days, and as it progressed, it was clear that social networks were playing a role. For example, retiree Terry Mays at first believed the Bank of England's guarantee was enough, but by Monday he had traveled to the London branch where he said, "I took some financial advice over the weekend, and I'm taking the money out to get peace of mind. We're relying on this money for our pension."[7]

This kind of person-to-person contact caused many people who would otherwise ignore the run to join in the frenzy. And the anxiety that spread was similar to the anxiety that spreads in mass psychogenic illness, which we discussed in chapter 2. Like MPI, bank runs take on a life of their own. Under the right conditions, what starts as aberrant behavior in just a few people can spread like wildfire in social networks.

These sentiments can spread not just among depositors but also among investors, creating a "banking contagion." As news of the Northern Rock run dominated the financial press, people wondered who was next, and panic soon started to spread to other firms. The bank Alliance & Leicester lost a third of its market value (1.2 billion pounds) shortly after the run on Northern Rock, and shares of other banks fell too. Soon there was a generalized fear that these other banks were in a similar situation and might need to make the same kind of announcement, which would set off a whole wave of bank runs. Fortunately, before things spiraled out of control, Alistair Darling, the Chancellor of the Exchequer, made an official statement that the British government and the Bank of England would guarantee

Northern Rock's deposits. The bank run came to an end, and the financial markets stabilized.

Of course, the story did not end there. The subprime crisis continued to unfold, and later in 2008, financial contagion swept through international markets. First, it hit institutions like Bear Stearns that were directly involved with the mortgage market (on the verge of bankruptcy, Bear Stearns was purchased by JPMorgan Chase for a paltry $2 per share). Then IndyMac Bank failed (the fourth largest bank failure in U.S. history), and soon the federal government had no choice but to take over Fannie Mae and Freddie Mac, two formerly private mortgage companies that guaranteed about half of the $12 trillion dollars in U.S. mortgages. A week later, the crisis spread to investment banks as cash-starved Merrill Lynch acquiesced to a buyout by Bank of America, and Lehman Brothers collapsed. Two days later, the crisis spread to insurance giant AIG, forcing the U.S. government to step in and lend the company $85 billion. As two other banks (Washington Mutual and Wachovia) failed, markets froze and banks stopped lending money. One investor joked darkly that the only safe investments were bottled water, bomb shelters, and a nice cubbyhole in his mattress. By October 2008, the U.S. government had agreed to fund a $700 billion bank rescue plan, but it was too little too late. The Dow Jones and S&P 500 stock market indices had fallen over 40 percent from their highs a year earlier, representing a stunning loss of $8.4 trillion.

The meltdown of 2008 shows how easy it is for panic to spread in financial networks. When one big company fails, others that are connected to it are also at risk. In fact, famed investor Warren Buffet, in his annual shareholder letter of 2009, characterized the cascading nature of the business failures as follows: "[Market] participants seeking to dodge troubles face the same problem as someone seeking to avoid venereal disease....It's not just whom you sleep with, but also whom they are sleeping with."[8] Hyperdyadic spread indeed.

As the losses continued, they led to a dramatic global slowdown in the economy that was the worst we have seen since the Great Depression. Thousands of people lost their homes, and millions more lost their jobs. Amid such a breakdown in trust between people and institutions, the only solution was government intervention. Once the government made it clear that it would intervene to prevent further failures, banks started to lend money again, and the markets began to stabilize. This has caused some experts to wonder if we could have prevented the problem by acting earlier.

Although many ties in financial networks are formal (for example, many affected firms had legal contracts with other firms that had failed), we should not underestimate the power of informal and personal ties. Wall Street has developed a vast culture that promotes social relationships between bankers and CEOs, salespeople and clients, and even between competing traders. These titans of industry and masters of the universe come into frequent contact on the phone, at business meetings, and during after-hours social events. And when they quit their jobs to move to new firms, they become connectors, linking everyone at their previous offices with everyone at their new ones. As a result, markets that move vast sums of money through the international financial system are run by tight-knit networks of traders where the major players often know one another so well that they can tell who they are trading with just by watching the pattern of bids and offers that appear on their computer screens. Traders could ignore this information, but they probably do not. When people they trust start selling, they may want to sell too. Although some company failures are to be expected during economic downturns, social networks can exacerbate the problem by spreading fear among the very people and institutions who must take risks to turn things around.

It might seem like the modern technological age has made us much more interdependent and therefore more susceptible to panics like these. However, the role of social networks in financial contagion is nothing new. Economists Morgan Kelly and Cormac O'Grada studied Irish depositors at a New York bank (Emigrant Industrial Savings Bank) during two panics in the 1850s.[9] They had an extraordinary amount of information about these depositors, including which parish in Ireland the depositors came from. Arguing plausibly that individuals from the same parish were likely to have known each other during this time, they used this information to construct social networks and to see whether socially close individuals corresponded in their decision to withdraw money during the panic. Kelly and O'Grada found that social networks were the single most important factor in explaining the closure of accounts during both panics, even more so than the size of the accounts or the length of time they had been opened. Thus, financial panics may result from the spread of emotions or information from person to person.

It is interesting that such economic phenomena are usually seen as aberrations. Traditional economists would say this behavior is not rational. After all, many people who stood in lines at Northern Rock to withdraw their money did not really think the bank would fail. Some even explicitly said so. But, spurred on by the motion of the herd, they blindly followed. In this way, social networks generate behavior that is not consistent with the simplified, idealized image of a rational buyer and seller picking a price to transact the sale of goods. And for many years, economists reacted to this inconsistency by ignoring the behavior altogether.

Bank runs are a classic example of how individually rational behavior can lead to communally irrational behavior. We are all capable of thinking with our heads, but our hearts keep in touch with the crowd, and sometimes this leads us to disaster. Social networks can make a problem worse because they make it possible for the first people who panic to influence many others (like the couple who decided to withdraw their money once they discovered their friends had). The wisdom of crowds can quickly turn to folly.

Where's George?

Social networks clearly play a big role in financial crises, but they also have an effect on everyday transactions. Have you ever wondered where the dollar bills you get from the cashier at the grocery store come from? Some are dog-eared and look like they have been through the washing machine at least a dozen times in forgotten pockets, the kind of bills that a soda machine just won't take no matter how many times you try to flatten them and feed them in just right. Bills like that have a history. They have passed from hand to hand in all kinds of transactions, from buying soda to paying the kid who mows the lawn to giving the grandchildren a present to buying drugs or sex. The dollars in your wallet have had a secret and varied life.

This life actually represents one path through the enormous social network we inhabit. If there were some way to see such paths, these endless exchanges in the whole human economy, then we might be able to better understand the ties that connect us. The flow of money depends on social-network ties but also defines them.

Lots of people are curious about where their money has been and where it is heading. Some people write their names on bills in the vain hope of receiving the bill back in the future. But in 1998, a database consultant from Brookline, Massachusetts, named Hank Eskin figured out a way to satisfy this curiosity. He started a website called Where's George? (WheresGeorge.com). The George he was looking for was George Washington, whose face first appeared on American dollar bills in 1869. Prior to the Internet, tracing the movement of currency in the fashion Eskin had in mind would have been impossible.

Eskin's website allows people to track a particular dollar bill by entering its unique serial number and the ZIP code where the bill was acquired into an online database. Anyone can record a bill to the database, and if that same bill has been previously entered, the website will show where the bill has been. These records are known as "hits." Visitors to the website can also leave notes about where they received the bill. And so it is possible see the specific paths that the dollars take from one person to another.

As of 2008, more than 133 million bills had been tracked, with a total value of over $729 million (the site accepts all denominations). One user, Gary Wattsburg, has entered almost a million of those bills himself, but the majority of the bills are reported by newcomers to the website.

Most bills are not reported more than once. But 11 percent reach two or more people. In fact, one of these bills was reported by fifteen different Where's George? users. This particular bill had a colorful life. It was first reported in 2002 in Dayton, Ohio, and soon traveled to Scottsville, Kentucky, where a user received it as a tip in his job at a drive-in restaurant. The bill crossed the border into Tennessee, where it was given as change at the Shell Food Mart in Chapel Hill, North Carolina, and at a country store in Halls Mill near Union-ville. The bill found its way to Texas, where one person received it as change in a McDonald's in the town of Keller. It then passed through an adult-entertainment part of the social network. The bill was given as change at a racetrack betting window at Lone Star Park in Grape-vine, and later it was found on the floor at the Penthouse Key Club in Dallas, a "sexually oriented business" that has been shut down a few times by the Dallas City Council for prostitution.[10] After a brief stint in Shreveport, Louisiana, the dollar returned to Texas in change at the Jack in the Box restaurant in Rockwall and later at Mr. K Food Mart in Irving. It apparently ended its spree in 2005 after passing through Panguitch, Utah, and later to Kincheloe and Rudyard, Michigan, where the last person to report on the bill wrote, "This bill is getting pretty old looking."

All told, the bill traveled at least four thousand miles in a little more than three years, averaging about 3.8 miles a day. No other bill has been so well tracked. But the entries at this website contain information about the "jumps" that countless bills have made, including the distance between origin and destination and the time it took for the bill to get from one place to another. These jumps can skip over people who did not report the bill. So, for example, the bill we just described was probably exchanged between many more than fifteen people. But never before had we known so much about where, when, and how money travels.

The flow of dollar bills through financial-contact networks resembles the flow of sexually transmitted diseases through sexual-contact networks. In these examples, the network can be deduced simply by what flows across it. This is good news for researchers, because both germs and money can be used to trace connections that might not otherwise be apparent. But inferred networks differ from fully observed networks. In an observed network (like a regular friendship network), we know all the connections, and we know who has the potential to transmit something to someone else, even if nothing is transmitted. For example, you might still be good friends with a friend from high school even though you have not had contact in years. In an inferred network, however, we only observe the realized interactions. Inferred networks are therefore incomplete pictures of social networks. So, for example, two people may have a sexual relationship but might never transmit a disease. The science of social networks often depends on the art of figuring out what kind of network to study and how to discern it.

SARS, Seagulls, and Sailors

In 2003, the world faced an epidemic of a new disease called SARS (severe acute respiratory syndrome). In the months following the appearance of the epidemic, many scientists became interested in the impact of social networks on the spread of disease. As we will discuss in chapter 8, over the centuries there has been a dramatic increase in how far people can travel, and the wider physical range of modern social networks has greatly increased the speed at which pathogens can spread. In fourteenth-century Europe, the plague (the Black Death) did not spread very quickly from town to town because people typically did not travel more than a few miles a day, and the people they interacted with usually lived nearby. Back then, it took more than three years for the plague to move from the southern part of Europe to its northern reaches, with an average speed of movement of two or three miles a day.[11] In comparison, one of the people involved in the 2003 SARS outbreak carried the infection eight thousand miles (from China to Canada) in a single day!

The urgency of the SARS epidemic prompted researchers to meet in Montreal to discuss the impact of social networks and human travel on disease. One problem in particular was the question of measurement: how is it possible to follow the movements and interactions of enough specific people in order to be able to build a statistical model to predict the spread of a pathogen? The answer to this question came soon after the conference. Dirk Brockmann, a researcher at the Max Planck Institute for Dynamics and Self-Organization, stopped in Vermont to visit a friend on his way back to Germany. Brockmann's friend, a carpenter by trade, was a fan of the Where's George? website, and he showed him how money could be traced from person to person and from place to place. Brockmann was intrigued. People carry dollar bills and then exchange them person-to-person in close contact, just like they carry and exchange viruses and bacteria. If the researchers could understand the movement of money, they just might be able to learn something about the spread of SARS, flu pandemics, and other deadly diseases.

Brockmann and his fellow researchers Lars Hufnagel and Theo Geisel soon contacted Hank Eskin at WheresGeorge.com to ask for the data. Eskin obliged, and soon the researchers were awash in the very data they were saying they so badly needed just weeks before.

As Hufnagel put it: "Since we can't track people with tracking devices, like we do animals, we needed to get data that provided us with millions of movements of individuals."[12] They would not have a record of every single transaction, but the sheer quantity of information they did have meant that they could describe general rules that apply even to the transactions they did not observe. The researchers reported their results in the prestigious journal *Nature* in early 2006.[13] Since then, scientists have begun to harvest still other sources of movement data, such as the traces left by cell phones, which we will discuss in chapter 8. Cell phone data allows researchers to study who people are connected to and where they are minute by minute for months at a time.

Brockmann and his colleagues discovered that the jumps bills make from one place to another obey a simple mathematical rule. A typical dollar bill is traded locally several times, moving only a few feet or a few miles between exchanges; but occasionally you take your wallet with you on a trip to a friend's wedding, a family gathering, or a business meeting halfway across the continent. And most of the time, money does not stay with you for long; it leaves your pocket shortly after it gets there. But sometimes, you lose track of money, and it stays with you for a very long time; you might forget about that twenty-dollar bill in your parka until you are happily reunited with it the following winter.

The overall pattern indicates two important features of human interaction. First, bills stay much closer to home for a much longer time than previous models of human movement had predicted. Our regular routine involves straying little and spending cash locally. Yet, when bills do jump from one place to another, the distance they jump is typically much longer than previous models of human behavior had predicted.

In fact, the jumps follow a mathematical pattern poetically called a *Lévy flight,* after the French mathematician Paul Pierre Lévy.

Imagine a seagull that is searching for food. It might find a nice spot by the seashore where it can catch crabs, and it will stay there for several hours chasing them in and out of the waves. But when the tide changes, it might then fly a long distance to reach its next feeding location. Lévy flights, with their pattern of many short jumps interspersed with a small number of very long jumps, are quite different from what are called *random walks*, where each jump is roughly the same size and in a random direction. For a typical random walk, instead of a foraging seagull, imagine a very drunk sailor. He starts by holding on to a lamppost. When he lets go, which way will he

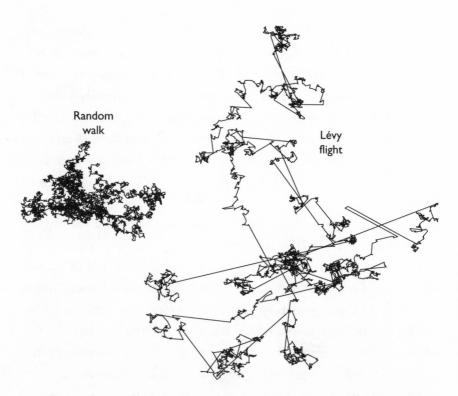

Random
walk

Lévy
flight

The random walk (left) shows five thousand steps of equal length in a random pattern of movement. In contrast, the Lévy flight (right) shows five thousand steps of varying length, sometimes with a "flight," in a random pattern of movement.

stumble? Left or right? Will he lunge forward or stagger backward? And if we leave him for a while and come back, where will he be?

Just like the bird, the sailor will appear to move randomly. But if we drew their paths, they would look very different, as shown in the illustration. Both would look like a tangle of spaghetti at first, criss-crossing more and more as time goes on. But at a certain point, the bird gives up on its current feeding ground and flies miles away to start a new search in a different location. The sailor, taking the same-sized step each time, can't do this (though if he is drunk enough, he might imagine that he can). As a result, we have very different predictions about how far the seagull and the sailor can travel in a certain amount of time. In the end, the sailor cannot really stray too far from the lamppost. Not so for the bird in Lévy flight. Because it can make the occasional long trip, it will be able to travel away from its starting point at a much faster pace over time.

Since the jumps of dollar bills look like a Lévy flight, the average speed of a dollar bill should be much faster than it would be if it were taking a random walk. However, Brockmann and his colleagues discovered that the movement of dollar bills from person to person followed a pattern that was somewhere between sailors and seagulls, traveling faster than a random walk but slower than a Lévy flight. To see why, they also studied the lengths of time between jumps, not just the distance. They found that, as with the pattern of distances, the pattern of times between dollar exchanges was dominated by many short intervals, but occasionally the intervals would be really long. Some dollars were traded frequently, while a rare few got stuck in the hands of an infrequent trader, in a bank vault, or with the socks lost in the laundry. This could help explain why the dollars would spread more slowly than expected in a social network where the movements of people followed a Lévy flight. And modeling both time and distance in the network of financial transactions helped researchers to better understand how often people come into physical contact and how fast a disease like SARS might spread.

Moody Markets

The famous mathematician Benoît Mandelbrot developed much of the mathematics used to describe Lévy flights. He used these new techniques to study price changes he had first observed in the early 1960s in the cotton market and other financial markets. Scholars had previously thought that price changes in these markets followed a normal bell-curve distribution, with many average-sized jumps and just a few jumps of moderate size occurring every now and then. But Mandelbrot showed that both small changes and large changes were much more common than expected. Like the foraging seagulls, markets tend to oscillate near a given price for a while and then jump to a new one.

There are lots of reasons why markets might make a long jump from one price to another, and our interconnection is one of them. Some pieces of information are so important that markets respond to them in seconds. For example, the government routinely releases statistics about economic growth, unemployment, the housing market, and inflation that can have a big effect on bond and stock prices. But another reason is that prices are not just an impartial estimate of the objective value of an item: prices also include expectations about how much other people value an item. The more people who think gold is a good investment, the higher the price will go. We do not simply decide for ourselves how much gold is worth; we also look at what others think gold is worth in order to decide. Our judgment about the value and desirability of goods is thus similar to our judgment about the value and desirability of sexual partners: it depends on how others perceive the object of affection in question. Social pressures can drive demand.

This makes markets much different from foraging seagulls. When a seagull eats a crab, it gets the same nutritional value from it no matter how many other seagulls wanted to eat it; a crab is just a crab. In

contrast, when a person buys gold, the profit he makes depends critically on the number of other people who also want to buy gold.

So what determines the number of people interested in buying
gold? Economists say markets are driven by supply and demand,
but where does the demand come from? In part, it comes from the
inherent value of an object. Gold can be used to make wedding rings,
royal crowns, foil for space capsules, and teeth. But demand is also
influenced by needs and expectations, and these can be strongly influenced by the needs and expectations of others to whom a person is
connected. Moreover, people may need to have confidence that they
are investing in something that others will want to buy in the future.
This gives markets an inherently subjective quality.

For example, you might be able to make $500 worth of stuff
with a one-ounce gold coin, but if you think that someone else in
the market is willing to buy it for $1,000, then you will probably
try to sell it for that price. And once you ask for $1,000, you send a
signal to everyone who sees your shiny gold coin for sale that you
think it is worth much more than $500. You might not get your ask-
ing price, but you might get more than $500. If you do, then the
increase in the price of gold sends a signal to other market participants. Increasing prices may convince some people that demand for
gold is on the rise, which could increase their confidence that others
would be willing to buy it at a higher price in the future. Like people
doing *La Ola* at sporting events, market investors take cues from
one another in synchrony, driving prices away from reality. It is just
this situation that results in "irrational exuberance" in stock markets, housing markets, even tulip markets (in seventeenth-century
Netherlands).[14]

Human social networks thus have economic moods. Nothing
makes the collective nature of these moods more obvious than the
language we use to describe changes in the economy. The economic
boom in the 1890s in Boston and New York gave rise to the decade's
moniker "the gay nineties," and we use equally evocative expressions

when we speak of economic downturns as "panics" and "depressions." As discussed in chapter 2, moods can spread from person to person to person, making the situation even worse than the objective situation in the economy warrants.

At this point, traditional economists may cry foul. After all, from Adam Smith on, the conventional perspective has been that markets are efficient: an "invisible hand" leads to the "correct" price for the good being traded. If too many people think the price is too high, it will fall because people will buy less. If too many people think the price is too low, it will rise because people will buy more. The most recent price reflects the best guess about where these expectations are in balance.

And, in fact, we have lots of examples where the market does a pretty good job of getting it right. One of the most prosaic comes from "Vox Populi" (Latin for "voice of the people"), a 1907 article in *Nature* by polymath statistician Francis Galton.[15] Galton visited the West of England Fat Stock and Poultry Exhibition, a county fair where there was a contest to guess the weight of a fattened ox. Participants had to pay six cents to guess, and the closest guesses won prizes. Galton managed to acquire the cards on which people had made their guesses, and he showed that most guesses were quite bad. However, when he ordered them from the lowest guess to the highest guess, he found that the median guess (1207 pounds) was extremely close to the actual weight of the ox (1198 pounds). Galton concluded, to his own surprise, that democratic decision making might not be as bad as previously thought. When faced with the challenge of identifying the correct weight of the ox, most individuals would get it wrong, but the group as a whole could get it right. If the ox had been for sale, the same thing would have happened with respect to its price, and the ox's true value could be determined.

More examples come from modern-day election-prediction markets, like the Iowa Electronic Markets and Intrade. In these markets, using real money, you can buy an outcome, and if that outcome occurs, you get paid. For example, for the 2008 election, you could buy contracts for Barack Obama, John McCain, and all the other candidates for the presidency. If you bought an Obama contract, you got paid the day after the election, but if you bought any of the others you did not, since Obama won. The price in these markets reflects the probability that people think the outcome will occur. So if an Obama contract that pays one dollar is priced at sixty cents, it means that the market expectation is that Obama has a 60 percent chance of winning. Scholars have compared market predictions with what actually happens, and they have shown that election markets predict outcomes better than other available methods, such as polling.[16] In fact, they are so successful that prediction markets are now commonly used inside large companies like Siemens, Google, General Electric, France Telecom, Yahoo, Hewlett-Packard, IBM, Intel, and Microsoft to aggregate information about production schedules and competitors. The employees make bets on what is going to happen. Such markets can even be used to predict the risk of terrorist attacks.[17]

While economists will point to markets like these to emphasize the triumph of the invisible hand, it is important to note that they are, in fact, special cases of group activities. In the fattened ox example, individual guesses were made independently. No doubt, some people discussed their guesses with friends at the fair, but the guesses were not made public like prices on the stock market. Moreover, the payoff was explicitly tied to an objectively verifiable event. The ox got on the scale, and a winner was determined. Similarly, in the prediction markets, an outcome occurred, and people got paid.

In contrast, stocks and houses are continuously traded until a company goes bankrupt or the house burns down. It is true that companies report profits at regular intervals, and these reports have an effect on perceptions of value. It is also true that the price of building a new home constrains how much someone is willing to pay for an existing home. However, the overall value of stocks and homes is highly dependent on what other people think they are worth. Competitive markets may operate via the invisible hand, but social networks can distort these markets, sometimes yielding an invisible slap in the face.

Although there is often wisdom in crowds, they also can go horribly awry when making a decision. The difference between these two extremes (say, an orderly election and a violent riot) has a lot to do with the path-specific motion of information through networks. Whether groups of people are able to reach a correct decision about something (the value of a product, the number of jelly beans in a jar, the weight of an ox) depends on whether decisions are made at the same time or sequentially. If a group of people is deciding on the price of an item and bid on it independently, then their average guess is probably a good indicator of its market value. However, if people make decisions in sequence and are aware of prior decisions, if information moves from one person to the next (as in the game of telephone), we can end up with the blind leading the blind. Once a critical mass of people make a decision, the rest of the group goes along, reasoning that others cannot all be wrong. Like the people in chapter 1 looking up at the window in New York City, they fall in line. So whether the wisdom of crowds can be trusted may depend on whether the members interact concurrently and independently or sequentially and interdependently.

Sociologists and physicists Matthew Salganik, Peter Dodds, and Duncan Watts studied this problem using an online music market.[18] They designed an experiment involving an online site they created that gave away downloadable songs. A total of 14,341 people came to a website featuring forty-eight songs. There were different "worlds," however, that visitors to the website could experience, and these worlds were created by the actions of previous users. Visitors could download songs from bands they had never heard before and evaluate their quality after listening to them. In one "world," subjects were able to see what previous participants thought of a song's quality, while in the other "world" they could not. The scientists found that in the world where song ratings were visible, the first person's rating influenced the whole trajectory of ratings for particular songs, keeping them high for a very long time. In other words, musical tastes are contagious. A tiny tweak in the sequence of social interactions when people make cultural choices can turn an average tune or a mediocre singer into a sensation.

This experiment documents the path dependency that can arise when people make decisions in sequence. There is no correct or true value of the songs in question. The value and the quality of each song depends on an idiosyncratic and essentially random process that gives rise to a particular sequence of people making choices. Because of our tendency to want what others want, and because of our inclination to see the choices of others as an efficient way to understand the world, our social networks can magnify what starts as essentially random variation. And these small variations can sometimes cause big differences in whether or not we can work together to solve problems.

Three Degrees of Information Flow

Residents in the high Andean community of Tigua Loma had beautiful latrines. As a Peace Corps volunteer in Ecuador, James visited Tigua Loma and worked in many communities where basic sanitation was a challenge and where preventable diseases like cholera were epidemic. Time after time, economic development agencies poured money into latrine projects. Each family spent hours digging pits, moving materials, and building walls. When the latrines were finally completed, the community celebrated with the engineers who had helped them through the entire process. However, in Tigua Loma, the latrines would often fall into disuse. Why? And why did some communities succeed in changing their behavior when others failed?

New technologies frequently offer enhanced quality of life. Even basic inventions like water pumps and latrines can dramatically improve the health and economic well-being of people in remote parts of the nonindustrialized world. Yet all too often, even when resources for these new technologies are available, they fail to catch on. Figuring out how and why people adopt new ideas and how they can spread from person to person to improve underdeveloped economies has been a driving force in the science of social networks since its inception. In fact, some of the earliest concerns had to do with how new ideas spread in a population. Development experts wanted to know how they could spread more efficient agricultural techniques from farmer to farmer. Public health officials wanted to know how new medical practices spread from doctor to doctor or from family to family. And for-profit firms wanted to know how recommendations to buy their products spread from consumer to consumer.

One feature of this early work was that it rarely included information about specific social ties between individuals. For example, sociologist Everett Rogers's seminal book, *Diffusion of Innovations,* treated technology in a population like a drop of blue dye in a glass of water.[19] He theorized that technology would diffuse slowly at first, then quickly, and then slowly again as it reached the entire population. However, recent research that takes social-network structure into account shows that it is not so simple. In particular, many ideas never take off at all, and the influence of any one decision in the network may be limited.

Something was in fact different about the structure of the social network in Tigua Loma. The people there were more suspicious of each other, there were fewer "*mingas*" (shared labor between households during the harvest), and there were fewer connections between the people. Local institutions that might have given people opportunities to form ties were less present in this community than in others nearby. The people in Tigua Loma had a problem. They didn't talk to one another.

Scholars have started to focus on network structure and how it affects information flow. In one study, researchers carefully examined the word-of-mouth network of recommendations for three piano teachers in Tempe, Arizona. The teachers did not advertise and therefore relied on their social networks to keep them in business. Most of the recommendations occurred between close friends who were directly connected, but the positive references would spread, often to people the original recommender did not know. In fact, fully 38 percent of the recommendations came from people who were three degrees removed from the piano teacher they were recommending (the teacher's friend's friend's friend). However, the paths tended to fizzle out after that, with less than 1 percent of the recommendations reaching people who were six degrees removed.[20] The great majority of pupils came from within three degrees of the teachers.

The next example hails from a completely different corner. The spread of information is obviously crucial to the process of invention. And while information does spread between inventors, here too the spread is limited. When inventors submit patent applications, they nearly always connect their ideas to the work of other innovators by citing other patents. Although there are several reasons for this, the principal one is that the inventors acquired information from another invention that was useful in their own.[21] In addition, many patents are filed by two or more people, so the patent filings can be used to establish the social network of inventors—who collaborates with whom. There are thus two crucial sets of information in the patent data: the network of ideas and the network of collaborations.

An examination of more than two million citations from one patent to another used these citations to identify the effect of social networks on the spread of ideas among inventors.[22] It showed that there was a strong probability that inventors with a direct collaboration tie would cite each other; in fact, they do so about four times more often than would be expected due to chance. But the effect extended further into the network. At two degrees of separation (two people did not collaborate directly but instead shared a common collaborator), they would be about 3.2 times more likely to cite each other, and at three degrees (a collaborator's collaborator's collaborator), they would be 2.7 times more likely. Beyond three degrees, the effect virtually disappears. Additional analysis showed that these relationships did not exist just because two inventors happened to work on similar designs. Instead, they were a direct result of the spread of information through the social network.

The Strength of Weak Ties

The principal idea underlying the diffusion of innovation is that information and influence tend to spread through close, deep connections. If we have an effect on people we do not know, it is because we take advantage of a series of strong ties. Like dominoes falling one by one, we can spread information to, or influence the behavior of, the next person, and that person does so in turn.

However, this idea neglects an important feature of human social networks. As we discussed in chapter 1, we tend to be clustered in tightly knit groups. Take any two of your friends at random, and the chance they are friends with each other is higher than 50 percent. As a result, the series of strong ties through which we might influence others is not like dominoes. Ties do not extend outward in straight lines like spokes on a wheel. Instead, these paths double back on themselves and spiral around like a tangled pile of spaghetti, weaving in and out of other paths that rarely ever leave the plate.

While this structure is good for reaching everyone in your group, and even for reinforcing your own behavior via feedback loops, it is very bad for reaching people outside the group. Stanford sociologist Mark Granovetter was one of the first people to recognize this difference. Others had dismissed "weak ties" and casual acquaintances as irrelevant to the spread of information. But Granovetter argued that these weaker connections frequently act as bridges from one group to another and therefore play a critical role. Strong ties may bind individuals together into groups, but weak ties bind groups together into the larger society and are crucial for the spread of information about the benefits of using latrines, the availability of good piano teachers, the existence of valuable information in other inventions, and much else besides.

Granovetter used a simple economic study to prove his point. He surveyed several technical, managerial, and professional workers in a Boston suburb who had recently relied on a personal contact to get a new job, and he asked them a simple question: "Prior to switching employers, how often did you see the person who helped you get the new job?" He found that only 17 percent responded "often," while 55 percent said "occasionally"; the remaining 28 percent said "rarely." Most workers found jobs via old college friends, past workmates, or previous employers. Contact with the person was sporadic, and very few had ever spent time with the contact outside the workplace. According to Granovetter: "Usually, such ties had not even been very strong when first forged.... Chance meetings or mutual friends operated to reactivate such ties. It is remarkable that people receive crucial information from individuals whose very existence they have forgotten."[23] In other words, most of his subjects had acquired their jobs by (nearly) relying on the kindness of strangers. These were distant friends or friends of friends who passed their names to an employer or who passed information about jobs to the prospective employee. People find jobs, in other words, in much the same way that they find sexual partners (as we saw in chapter 3)—by searching their social network beyond their immediate ties.

Weak ties are thus a rich source of new information that we tap when we are trying to improve our lot. And we seem to do this intuitively, even though we do not know the structure of our own network or consciously think about the problem in the way Granovetter proposed. In fact, people frequently rely on weak ties to search large networks for useful information, as the study of global e-mail forwarding outlined in chapter 1 showed. People frequently relied on socially distant friends to accomplish this task. Since information flows freely within a close circle of friends, it is likely that people know more or less everything that their close friends know. Therefore, your immediate relatives and friends, for instance, would be unlikely to know something you do not about how to reach a person in Indonesia. But move socially farther away, and there is less overlap in experience and information. We might trust socially distant people less, but the information and contacts they have may be intrinsically more valuable because we cannot access them ourselves.

One implication of this is that people who have many weak ties will be frequently sought out for advice or offered opportunities in exchange for their information or access. In other words, people who act as bridges between groups can become central to the overall network and so are more likely to be rewarded financially and otherwise.

The other implication is that we sometimes leapfrog over the natural boundaries of the network when we are intentionally trying to search it for information and opportunities. The flow of influence may stop at three degrees, but it appears that we often start our search for information two or three degrees away in order to make sure we are learning something new. We do this in everyday life, whether searching for a job, an idea, or a new piano teacher, and it is this region in the network, just beyond our social horizon, that has a critical impact on our own economic fortunes.

Good Ol' Boys Through the Ages

To make it easier to think about how networks affect economic outcomes, it is often convenient to assume that our ties to others are fixed. But, as in the case of sexual networks (whose structure unfolds over time because people usually acquire partners sequentially) and loneliness networks (where connections can form and break depending on a person's characteristics), networks are not static; they are dynamic. The flow of money, information, and influence means that we affect our friends and our friends' friends, and in the process the network takes on a life of its own, changing shape as time goes by. If money makes the world go round, it does so not because it passively accepts the network. Wealthy individuals and big businesses shape their networks according to their financial and economic goals, and in turn, the shape of their networks has a big impact on whether they can achieve those goals. The good ol' boys circle together and take care of their own.

Some of the earliest evidence of attempts to mold networks comes from the Renaissance. Cosimo de' Medici rose to power in fifteenth-century Florence and headed up a coalition of families and partisans that consolidated the emerging banking system of Europe and then ruled northern Italy for three centuries. John Padgett, a political scientist at the Santa Fe Institute and the University of Chicago, has collected an enormous amount of information about the Medicis and other Florentine families during this time period and shown that dramatic changes in social networks deeply influenced our modern capitalist and democratic societies.

The growth of trade with Asia caused some families to suddenly become wealthy, upsetting a feudal social network that was extremely hierarchical and disconnected between groups. New-money families started competing with old-money families for social control, and to do so, they intermarried with, and gave power to, tradesmen and guilds (who were increasingly important with the rise of commerce) and vied for their alliance. At the center of this new social network was the Medici party, which spanned many of the previously disconnected groups. As a result, the Medicis were able to conquer once and for all the oligarchs who previously ruled Florence. In the last battle between the two sides on September 26, 1433, Rinaldo Albizzi, the leader of the oligarchs, tried to organize his supporters to attack city hall. But only a few showed up at a time, and their lack of enthusiasm caused them to drift away before they ever reached a critical mass. In contrast, the Medicis organized a massive preemptive response that gathered all their supporters at the Palazzo Vecchio. As a result, no military battle took place—the outcome was obvious, and the oligarchs quietly drifted into exile. The result of this change in the social-network structure (away from oligarchy) was reduced social control, and with it came new institutions that would democratize Florence and later other parts of Italy and the world. This convergence of money and open political systems created a big bang in the arts and sciences that has had an enduring impact to the present day.[24]

Similar processes are at work in modern corporations. Although today corporations rarely seal deals through intermarriage, they do share executives on their boards of directors. Some of these are celebrities—Bill Clinton sat on at least twelve boards at one point—but most are genuine businessmen who have typically served the same industry for a number of years.[25] Directors create network links between the multiple firms they serve, and they can easily pass information among them. This increases the chances for collusion and market manipulation and has been a source of congressional investigations for over a century.

One classic study of the eight hundred firms with the largest market capitalization (i.e., the highest total value of all their stock) found that bank boards were particularly well connected to the largest businesses, and these businesses were themselves strongly connected to other businesses in the economy, making banks the most central actors in the network.[26] This is exactly the pattern we might expect to find if we thought banks were trying to use the social network of board directorships to exert control over the most powerful players in the economy or to tap the flow of information about industries. However, since board meetings and discussions between directors are private, it has been very difficult to verify whether the network actually influences a firm's decisions.

One way to tackle this would be to examine a behavior that all boards engage in that is public and trackable, namely, political contributions. One would expect two firms with similar interests or located in the same part of the country to donate to the same political candidates. Yet, even accounting for this, increasing the number of mutual directors between two firms tends to increase the similarity in the profile of campaign contributions.[27] This suggests that increasing the social ties between large corporations helps them to synchronize their behavior.

Social networks also affect the way businesses exchange goods with one another. Overly simplistic economic theories of markets typically assume that firms will sell to the highest bidder and buy from the cheapest seller, regardless of the personal histories of those involved. However, real-world interactions are often based on personal relationships between businesses that are embedded (strongly connected) in stable networks of trust and reciprocity.

Sociologist Brian Uzzi, a professor at Northwestern whose mother worked as a dressmaker in New York, had personally observed how some businesses in the apparel industry were embedded while others were not. He conducted interviews at several of these firms and found that embedded firms were more likely to survive than those that did not rely on their personal networks to decide with whom to trade.[28] But he also found that too much embeddedness can be a bad thing. An unconditional commitment to a particular business partner (a strong tie) can be disastrous if it causes a firm to completely ignore opportunities with other firms (weak ties). Thus, there is a trade-off between building stable relationships with a certain group of partners and being willing to leave those relationships when changes in the market cause them to lose viability. It is important to have a mix of strong and weak ties, and hitting the sweet spot is key.

Networking Creativity

Uzzi extended his insight from dressmakers to a little-studied corner of the corporate world.[29] From *Cats* to *Spamalot,* Broadway musicals have been big business for decades, but investors usually have to follow their gut when they decide to back one show or another. *Bye Bye Birdie* starring Dick Van Dyke ran for 607 nights on Broadway and was a smashing success, but *Bring Back Birdie* was a flop and closed after just four. What was the difference? Why do some shows succeed and others fail?

Uzzi believed the social networks formed by the musical production companies played an important role, so he and Janet Spiro studied collaborations between the producers of 321 musicals that premiered on Broadway between 1945 and 1989. In particular, they were interested in whether collaborators formed "small-world" networks like those identified by Duncan Watts and Steven Strogatz in their seminal 1998 *Nature* article.[30] The idea underlying small-world networks is that they exhibit two important features: low average path length (people can easily reach others in the network through a small number of intermediaries, as Stanley Milgram's Nebraska mail experiment illustrated) and high transitivity (most of a person's friends are friends with one another). Watts and Strogatz showed that you could put everyone on a highly structured network (like a ring or a grid where neighbors are only connected to each other) and then just add a few random connections to turn it into a small-world network with low average path length. The result was a highly ordered network with lots of cliques (groups in which everyone is connected to everyone else) but also with many ways that information can pass between these cliques from person to person to person.

Uzzi found that teams made up of individuals who had never before worked together fared poorly, greatly increasing the chance of a flop. These networks were not well connected and contained mostly weak ties. At the other extreme, groups made up of individuals who had all worked together previously also tended to create musicals that were unsuccessful. Because these groups lacked creative input from the outside, they tended to rehash the same ideas that they used the first time they worked together. In between, however, Uzzi once again found a sweet spot that combines the diversity of new team members with the stability of previously formed relationships. The networks that best exhibited the small-world property were those that had the greatest success.

Production company networks with a mix of weak and strong ties allowed easy communication but also fostered greater creativity because of the ideas of new members of the group and the synergies they created. Thus, the structure of the network appears to have a strong effect on both financial and critical success.

Making better musicals might not be at the top of your list of world problems, but knowing how to spur creativity in teams has much broader applications. Uzzi has also studied human achievement and how it relates to social networks. Previous perspectives on scientific discovery, for instance, have stressed individual genius as the explanation for outstanding achievement, but over the course of the twentieth century, discovery and innovation increasingly came to be properties of groups rather than of individuals. Of course, innovation rarely, if ever, arises without input from others, as we saw with the inventor networks. Breakthroughs are created in collaborative circles, and networks can amplify talent (we have certainly seen this in our own experience, finding that complementary skills and knowledge enrich our joint work, making the whole greater than the sum of the parts). The empirical question is how to show whether individuals do better when they are part of teams than they would do if they acted alone.

To study this problem, Uzzi used citations as a marker for the "best" scientific work. In the scientific world, citation is a form of praise or at least attention. Uzzi collected data on 21 million scientific papers published worldwide between 1945 and 2005 and also 1.9 million patent filings from a fifteen-year period. He then compared the papers written by individuals to the papers written by teams. Using citation as a measure of quality, Uzzi found that, on average, team efforts were judged to be better and more important science than efforts by individuals.

Uzzi also evaluated whether there was any truth to what many academics know informally as the "thirty-foot rule." This rule states that people collaborate only with others within thirty feet of them. But as we saw in the case of sexual partners, where people shift from finding partners "in the neighborhood" to finding them through their network, and as we saw in the case of obesity where social-network connections were more important than geographic connections, physical distance is becoming less of a constraint on scientific collaboration. Studying 4.2 million papers published from 1975 to 2005, Uzzi found that collaborative teams involving researchers at different universities are increasing relative to teams that are all from the same university. This trend has to do with a greater focus on specialization, and it surely has been spurred in part by globalization. But what is increasingly clear is that scientific collaboration works best in small-world forms of organization that make it easy to work with a mix of people from different places.

Color Coordinated

Although Uzzi's studies show an association between certain network shapes or structures and collaboration, it is hard to know if networks are causing people to collaborate differently or if people who are more likely to collaborate just happen to form certain kinds of networks. For this reason, computer scientist Michael Kearns and his fellow investigators at the University of Pennsylvania decided to create an experiment to see how social networks constructed in a laboratory influenced collaboration. They took students and arrayed them into networks of thirty-eight people that had different structures, such as those shown in plate 5.[31] The investigators gave students at each position in the network a single choice: what color do you want to be? And they were also given a single goal: choose a color that is different from the colors chosen by the people you are connected to.

Students were seated at computer terminals showing the colors chosen by their neighbors (they could not see the whole network), given a menu of colors, and told to pick a color different from their immediate network neighbors. They could change their color at any moment. And they were timed. If the group reached a solution in the time allotted so that every individual had a color different from his neighbors, then they earned some money.

So how did they do? It turns out that the structure of the network indeed had a big effect on their ability to solve the problem. Ring networks (like A–D in plate 5) were easier to solve than the more jumbled networks. And, counterintuitively, the more neighbors the average person had in the network, the faster the group as a whole arrived at a solution. The average time it took for the thirty-eight individuals to finish declined from 144 seconds (network A) to 121 (network B) to 66 (network C) to 41 (network D). The more complicated networks took still longer for people to solve (network E took 220 seconds and network F, 155).

The contrast between networks D and E is especially telling. The people in these two networks faced very similar circumstances, with about the same average number of neighbors and about the same average degree of separation between any two people in the network. Crucially, students in these experiments could not tell what kind of network they were in; all they could see was their immediate neighbors. Yet, network E took more than five times as long to solve as network D. So small differences in the overall patterns of connection in the network can matter a great deal to the performance of the group.

The lesson for people trying to coordinate efforts to solve economic problems is that it may be valuable to create explicit links in networks or to organize them in a way consistent with the task at hand. For example, the $787 billion economic stimulus legislation enacted in 2009 provided funds for thousands of local, state, and federal agencies that were all supposed to spend the money as quickly as possible. And to avoid wasteful duplication, they were all supposed to spend the money on different projects. The Kearns experiment suggests that, for projects like these, the government should create structured channels of communication between agencies in addition to whatever informal channels may already exist. In other words, the government should foster small-world connections.

But sometimes actors do not always agree on the goals they are trying to achieve. Consider the debacle in the federal effort to get aid to victims of Hurricane Katrina. Federal authorities wanted to evacuate people from New Orleans, but local police in Gretna (a town near New Orleans) feared being overrun and prevented evacuees from leaving the city.

Kearns and his fellow investigators wanted to know how networks affected decision making in exactly these kinds of situations, when people have different incentives but still must work together. They conducted another set of laboratory experiments in which people embedded in networks of varying structure attempted to reach a global consensus (everyone must be the same color).[32] And this time, the researchers created tension in the group's goals. Half of the subjects were told that they would earn an extra 50¢ if everyone was colored red and the other half were told they would earn an extra 50¢ if everyone was colored blue. As in the previous coloring experiments, if consensus was not reached by a certain time, no one would get paid. If people were stubborn, holding out for the extra reward, then no one would get anything. So some people had to yield.

Again, the speed at which consensus was reached varied according to the network structure. In networks where some people had many more neighbors than others, those with the most neighbors were able to drive the entire network to their preferred color. The investigators called this the *minority-power effect*. A small group of influentially positioned individuals can consistently get their way. On the other hand, such a group can also facilitate global unity and prevent the outcome where no one gets anything. So although social networks may help us do what we could not do on our own, they also often give more power to people who are well connected. And as a result, those with the most connections often reap the highest rewards.

Your Friends Are Worth Something

While elites like corporate directors clearly benefit from shaping social networks to suit their needs, it is less clear whether these benefits reach other levels of society. If anything, social networks might be seen as an explanation for why the rich are getting richer, and why economic inequality continues to rise. The logic is simple: if you are rich, you can attract more friends, and if you have more friends, you can find more ways to become rich. And recent changes in technology might make the problem worse. When it is easier to search and navigate social networks, the positive-feedback loop between social connections and success could create a social magnifier that concentrates even more power and wealth in the hands of those who already had it.

Fortunately, the millions of poor around the world are not completely out of luck. Over the past thirty years, there has been an important movement to use social networks to fight inequality and improve the lot of the worse-off by giving them access to something they never had before: credit. Although it may be hard to believe in the United States, where we get sent unsolicited credit cards in the mail nearly every day, many people in the rest of the world cannot borrow even a dollar. And the main reason they cannot is that they have no collateral; they do not own land or property, and what few things they possess have such limited value that traditional lenders do not consider using them as a guarantee.

Traditional banks around the world overlooked, however, a source of collateral that even the most destitute have: their friends and family. Social networks are ubiquitous, and, as it turns out, they can be used to successfully guarantee a loan. Bangladeshi economist Muhammad Yunus is credited for having this insight, which he originally developed when visiting poor villages near Chittagong University where he worked. When Yunus learned that women in the village of Jobra were being gouged by local moneylenders to pay for the bamboo they turned into furniture, he agreed to lend them money himself. What was the staggering sum these forty-two women asked for? About twenty-seven U.S. dollars. Less than a dollar each. The new microcredit market was born.

Sensing the need for this kind of loan throughout the country, Yunus approached a bank and became a guarantor for loans the bank would make to the villages, since it did not loan money to people without assets. Amazingly, the repayment rate for the loans he made actually exceeded the repayment rate the bank typically enjoyed. Yunus would go on to found the Grameen Bank in Bangladesh, which pioneered the microfinance loan.

One of the most important features of these very small loans is that they are given to groups, rather than to individuals, to help them start small businesses or make other investments that will help them escape poverty (like paying for their children's school or paying off high-interest loans from local moneylenders). In essence, individuals use their friends and family as social collateral to assure the bank that they will repay the loan. This makes these high-risk loans feasible because it dramatically reduces the probability of default. Social networks help distribute risk and help groups cope more effectively with unexpected events like a drought or a death in the family. But, most generally, this is a way to monetize social-network ties. The bank typically requires five people to form a group, and if each of the five successfully passes a test after a week of training in business skills, then individuals in the group are eligible to apply for loans. Loans are made to two people first, and if those are repaid, then the next two people can apply, and finally if those are repaid, then the fifth group member can apply.

Yunus attributes the success of the Grameen Bank model to features of the social network: "Subtle and at times not-so-subtle peer pressure keeps each group member in line." The bank also refrains from creating groups artificially since "solidarity [will] be stronger if the groups [come] into being by themselves." Up to eight groups are tied together and administered in centers where initial loan applications are screened by an elected member. This small-world design is exactly what Brian Uzzi found among dressmakers, on Broadway, and in academia. The Grameen Bank fosters strong ties within groups that optimize trust and then connects them via weaker ties to members of other groups to optimize their ability to find creative solutions when problems arise. According to Yunus: "A sense of intergroup and intragroup competition also encourages each member to be an achiever."[33]

Another important network feature is the bank's almost exclusive focus on lending to women. On the one hand, this makes sense since women may have more social collateral than men. But lending to women has the added bonus of multiplying the benefits of the loan since women are much more likely than men to invest in improving the lives of children via schooling and improved health services. Women are also more likely to invest in their husbands than men are in their wives.

Since the founding of the Grameen Bank, microloans have been shown to reduce poverty, even among the poorest of the poor, and their success at grass-roots development has spawned similar programs in more than a hundred other nations. Even the industrialized world is starting to use programs like these for college students and other low-income individuals. It is interesting to see how an innovation from Bangladesh that was built on a deep understanding of the natural advantages of social networks has itself spread. The microfinance movement has generated so much interest worldwide that Wall Street now packages the loans and sells them as bonds just like mortgages or other common securities. The Nobel Foundation recognized the efforts of the Grameen Bank and Muhammad Yunus "for their efforts to create economic and social development from below" by awarding Yunus the Nobel Peace Prize for 2006.

Similar institutions that capitalize on social ties have emerged throughout history. For example, rotating credit associations, also known as solidarity groups or money-go-rounds, are composed of people who voluntarily assemble into a group that meets periodically to contribute to a fund that is then given in whole or in part to one of the contributors in rotation. These associations are typically self-organizing; they do not rely on formal institutions and typically lack a leader. These types of associations are found all over the world, from Korea to China to Japan to Pakistan to India to Nigeria to Cameroon, and they are often used by immigrant groups in the United States to pool capital for entrepreneurial activities (especially since immigrants are often cut off from the formal banking sector). Similar groups were found among working girls in England in the nineteenth century. And traditions of barn raising among nineteenth-century frontier farmers in the United States were a variant: people would band together to take turns, perhaps on the first Sunday of each month, to build a barn for each person.

Anthropologist Clifford Geertz outlined in 1962 what may have been the first academic description of these institutions, noting that their origins may be related to the tradition of "rotating feasts" in which each person in a small group agrees to host a feast in turn. In the Indonesian setting where Geertz did his original fieldwork, and in most other settings where rotating credit associations were found in traditional cultures, the local people often saw these associations as less about serving economic objectives and more as serving social and symbolic functions. "This association strengthens our village solidarity and our communal harmony," they might say.[34] A bewildering array of such traditional institutions can exist, and some may even involve complicated procedures for charging interest or determining the order of receipt of the funds.[35] But what they all have in common is that social connections function to prevent defection after a person has received the pot. The dollars move from person to person, across established social ties, and everyone knows where George is.

It remains to be seen whether we can harness the power of social networks to improve the lives of the poor as fast as we are improving the lives of the rich. However, we feel optimistic that networks can be used to reduce inequality, both directly via loans and sound economic policies that cope with moody markets, and indirectly via improved physical health and mental well-being. The main unsolved question is not about whether we have the ability to use social networks this way, but whether we will. In other words, how do networks affect our capacity to govern ourselves and to achieve our goals of spreading well-being?

CHAPTER 6

Politically Connected

I n his election-night acceptance speech on November 4, 2008, Barack Obama said, "I was never the likeliest candidate for this office. We didn't start with much money or many endorsements. Our campaign...was built by working men and women who dug into what little savings they had to give five dollars and ten dollars and twenty dollars to this cause." In fact, Obama's team shattered records in fund-raising. By the end of the campaign, he had received $600 million in contributions from more than three million people. In hindsight, Obama's campaign was described as a perfectly run operation that made few, if any, mistakes. But how did he get people on board before the public perception that things were going well? How did he persuade so many previously uninvolved people to donate money to him and to vote for him, especially those who in the past believed their vote did not count?

In no small measure, Obama succeeded because these "working men and women" felt connected. Obama's campaign was a historical milestone in all kinds of ways, but the most revolutionary may not have been its fund-raising. Many have commented on Obama's remarkable ability to connect with voters, but even more impressive was his ability to connect voters to each other.

The 2008 presidential election saw a dramatic increase in the use of the Internet across all political campaigns, but Obama's in particular took advantage of the power of online social networks and social (person-to-person) media. In fact, Obama's use of the Web evoked comparisons with John F. Kennedy's use of television to win the race for the presidency in 1960. Both men forever changed the face of politics through their use of new technology, forcing friends and foes alike to adopt their methods of reaching out to the masses.

Lacking an established base, Obama realized very early that the Internet would be important. In 2004, Howard Dean had used the Internet to challenge traditional candidates, but online social networking had not yet arrived. Dean's campaign raised a lot of money but failed to mobilize supporters because they were not yet connected to one another. Obama recruited two very talented people to run his online campaign, Joe Rospars, a veteran of Howard Dean's campaign, and Chris Hughes, one of the cofounders of Facebook.

Hughes built the social networking site My.BarackObama.com, which logged 1.5 million accounts at its peak. Users could discuss the candidate, donate money, and, crucially, organize real-world social activity. Over the course of the campaign, more than 150,000 campaign-related events would take place in all fifty states. Meanwhile, supporters online formed 35,000 groups based on geographic proximity, affiliation with specific issues, and shared pop-cultural interests. Users of iPhones could download an application that made it easy to call friends to encourage them to vote or contribute to the campaign. The application organized telephone contacts by order of importance, listing first those friends in swing states where the election was expected to be close. And in the critical last week of the race, the Obama campaign organized more than a thousand phone-banking events to get out the vote.

According to the Pew Internet and American Life Project, all this activity made a difference.[1] Obama supporters were more likely than Clinton supporters to mobilize their friends and family members by signing online petitions and forwarding political commentaries via text, e-mail, or online social networks. In part, this was because younger people were more likely to support Obama, but even with members of the same age group, Obama supporters tended to reach out to their own networks far more often than Clinton supporters did. And the gap between supporters of Obama and McCain was even larger; it would eventually carry Obama to victory.

Your Vote Doesn't Count

Many people—whether Republican, Democrat, or Independent—felt inspired by their participation in the 2008 campaign. And many more encouraged their friends and families to vote because "it's the right thing to do." But this behavior is somewhat puzzling. Although adults in most democratic countries have the right to vote, each of these votes is just one of millions of others. Politicians frequently tell their supporters "every vote counts," and people usually say they vote in order to help their candidate win. But under what circumstances will a vote actually do that? This basic question has led to a series of investigations by brilliant social scientists, each building on the work of previous thinkers, but all leading, alas, to the same conclusion. Rationally speaking, each vote *doesn't* count. The reason we vote, it turns out, has a lot to do with our embeddedness in groups and with the power of our social networks.

In 1956, Anthony Downs, an economics graduate student at Stanford, decided to apply the science of "rationality" to the study of politics.[2] He did not mean this as an oxymoron. The word *rationality* takes on a very specific meaning here, and it is not the opposite of crazy. *Rationality* means three simple things. First, rational people have preferences and know them. So you prefer an orange to an apple, a dollar to a penny, or a Democrat to a Republican. Or you may be indifferent. The point is that you can compare two things, and you know which one you like better or you know you like or dislike them equally. Second, rational people's choices are consistent. If you would rather have the orange instead of the apple, and the apple instead of the pear, then you will not choose the pear over the orange. Consistency is analogous to transitivity in algebra: if $A > B$ and $B > C$, then it must be true that $A > C$. And third, rational people are goal oriented. Once we know what we want, we try to get it.

Downs wanted to see whether voting could be understood as rational and, if so, under what circumstances. He noticed that politics in the United States was often about two choices, not more. Vote for the Democrat or the Republican. Lower taxes or raise them. Veto the bill or sign it. In fact, our government is replete with formal procedures that reduce a wide range of choices to just two. Downs assumed that voters would focus on one of the alternatives (say Barack Obama) and think carefully about everything that would happen if this alternative were chosen. They would then assign a value to this outcome that described the benefit. In other words, they would try to answer the question, how useful would an Obama presidency be to me personally? Next they would think carefully about the other alternative (say John McCain) and assign a value to that future outcome. Each voter would then vote for the alternative that yielded the greater value for themselves.

But voting is not a required act in the United States, nor is it in most countries in the world. What makes someone decide to even bother showing up at the polls? Downs noted that voters would also take into account the costs of voting. We might need to take time away from our workday or leisure activities to go to the polls. For example, in the 2004 U.S. presidential elections, some voters in Ohio waited in the rain for hours to cast their ballots. It might also be personally costly to spend time collecting information about the election so as to know whom to vote for.

Taking costs and benefits into account, each person would then decide whether to vote. If a voter thought she would benefit equally from both alternatives, she might decide not to pay the costs of voting and stay home. Downs called this *rational abstention*—it makes sense for some people not to participate because they literally think, "There's not a dime's worth of difference between the two." Conversely, people who believe one alternative is much better than the other are likely to care a lot more about the outcome and would therefore be more likely to stand up and be counted, even if the costs of voting are very high. Those Ohio voters soaked to the bone are just one example of such highly motivated individuals.

But does this really explain why people vote, especially when they may also think that they cannot influence the outcome? Do they simply calculate the benefits and costs and make a choice?

Actually, it's more complicated than that. William Riker, a tremendously influential political scientist at the University of Rochester in the 1960s and 1970s, pointed out that Downs had overlooked the important fact that not just one voter makes this decision but millions.[3] Thus, to determine the value of voting, we need to decide not just who we like better but also the probability that our action—our vote—will help that person win. Calculating this probability may seem like an impossible task because there are so many possible outcomes. Obama could beat McCain by 3 million votes. Or he could beat him by 2,999,999, or he could lose to McCain by 1,345,267. Or…There are literally millions of possible outcomes.

Of course, there is only one circumstance in which an individual vote matters. And that is when we expect an exact tie. To see why this is true, ask yourself what you would do if you could look into a crystal ball and see that Obama would win the election by 3 million votes. What effect would your vote have on the outcome? Absolutely none. You could change the margin to 2,999,999 or to 3,000,001, and either way Obama still wins. Notice that the same reasoning is true even for very close elections. No doubt some citizens of Florida felt regret about not voting in 2000 when they learned that George W. Bush had won the state (and therefore the whole election) by 537 votes. But even here, the best a single voter could do would be to change the margin to 536 or to 538, neither of which would have changed the outcome.

So what is the probability of an exact tie? One way of looking at this is to assume that any outcome is equally possible. Suppose 100 million people vote for Obama or McCain. McCain could win 100 million to 0. Or he could win 99,999,999 to 1. Or he could win 99,999,998 to 2. You get the idea. Counting all these up, there are 100 million different possible outcomes, and only one will be a tie. Because roughly 100 million people vote in U.S. presidential elections, that would mean that the probability of a tie is about 1 in 100 million.[4]

The exact probability is obviously much more complicated than this, as it is unlikely that Obama or McCain would win every single vote cast. Close elections are probably more likely than landslides. So instead of theorizing about the probability of a tie, we could study lots and lots of real elections to see how often a tie happens. In one survey of 16,577 U.S. elections for the House and Senate over the past hundred years, not one of them yielded a tie.[5] The closest was an election for the representative for New York's 36th congressional district in 1910, when the Democratic candidate won by a single vote, 20,685 to 20,684. However, a subsequent recount in that election found a mathematical error that greatly increased the margin, meaning there are actually no examples of single-vote wins.

In this survey of elections, the average number of voters per election was about one hundred thousand. This is far fewer than the millions who turn out for a national election, and therefore we would expect the odds of a tie in a national election to be even lower. However, calculating this probability is not easy. U.S. presidential elections are complicated because they are not decided by the popular vote. Instead, each state has a number of electors it sends to an electoral college to choose the president. Bigger states get more electors, and most states award all their electors to the candidate who wins the popular vote in their state. As a result, it is possible to win a few big states by a narrow margin and gain the presidency by winning the electoral college vote while losing the popular vote (as George W. Bush did in 2000). Taking all of these complications into account in one big statistical model, political scientists Andrew Gelman, Gary King, and John Boscardin used real data from one hundred years of presidential elections to model the vote within each state and the effect this would have on electoral college votes.[6] Their model showed that the chance of a tie in any state changing the state's electoral college vote and hence the outcome was about 1 in 10 million.

So let's go back to the original question posed by Anthony Downs. Suppose you were deciding whether to vote in the 2008 election. When, given all this, does it make rational sense to vote?

First, you have to put a value on the difference between a McCain presidency and an Obama presidency. One way to arrive at this value is to ask yourself, How much would I be willing to pay to be the only person who gets to choose whether McCain or Obama is president? You can go to the bank and withdraw any amount you like. How much would you hand over to be the kingmaker, the one person who chooses who runs the country for the next four years? One dollar? Ten dollars? One million dollars? When undergraduates answer this question, they usually give amounts of less than $10, which is astonishing since this is probably the greatest value anyone could get for a $10 purchase. However, for the sake of argument, let's say you think it is a very important decision and you are willing to spend $1,000 of your own money to be the only person who chooses the next president of the United States.

Second, you have to account for the fact that, by voting, you get the opportunity to determine the election's outcome only when there

Plate 1. A network of 1,020 connected friends, spouses, and siblings from the Framingham Heart Study in the year 2000. Each node represents a person, and its shape indicates gender (circles are female; squares are male). Lines between nodes indicate relationships (black for siblings, red for friends and spouses). Node color indicates how happy each person is, with blue shades indicating the least happy, and yellow shades the most happy; shades of green are intermediate. Unhappy people and happy people tend to cluster in separate groups. In addition, unhappy people are more likely to be at the periphery of the network.

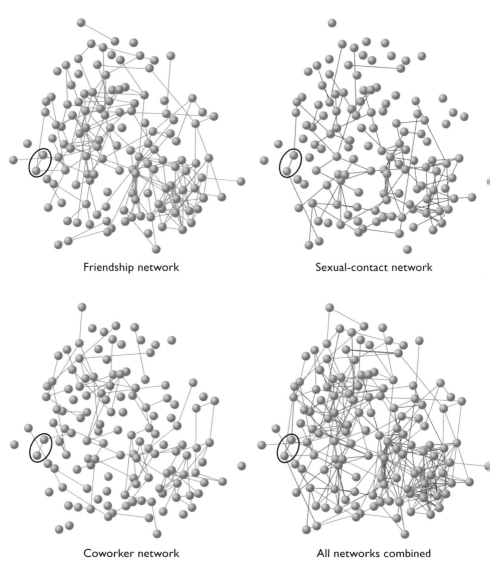

Plate 2. Different kinds of networks can overlap as shown here in a hypothetical network of 130 people. The network of friends (top left) is also the primary source of potential partners in the sexual-contact network (top right). Other networks, like those on the job (bottom left), also may be a source for friendship and potential sexual partners. The bottom-right network is multiplex, with several different kinds of relationships combining to contribute to the full social network. Some people have multiplex relationships with one another (for example, two people could be friends, coworkers, *and* sexual partners as indicated by the circled pair).

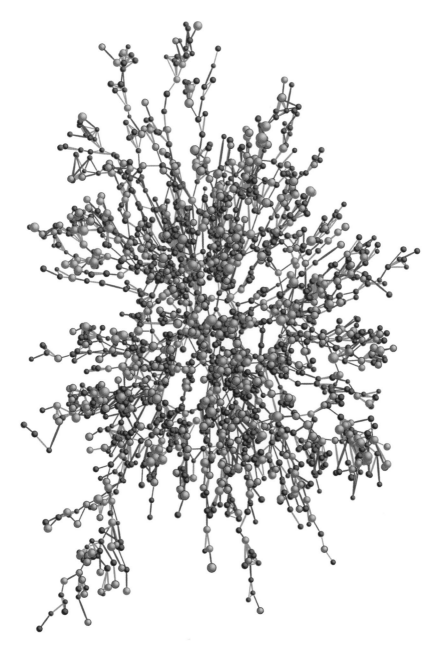

Plate 3. A network of 2,200 people from the Framingham Heart Study in the year 2000. Node borders indicate gender (red for female, blue for male); node colors indicate obesity (yellow for body mass index [BMI] of 30 or more, green for BMI less than 30); node sizes are proportional to BMI; and tie colors indicate the kind of relationship (purple for friend or spouse, orange for family). Clustering of obese and nonobese individuals can be observed within particular locations in the network.

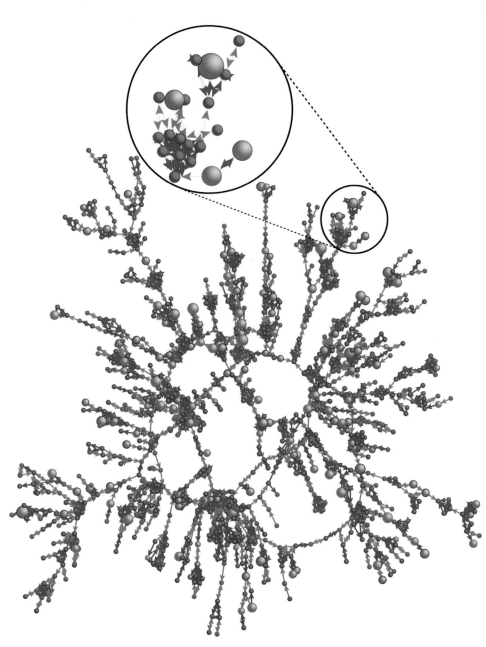

Plate 4. A random sample of one thousand subjects in the Framingham Heart Study social network in the year 2001. Node borders indicate gender (red for female, blue for male); node colors indicate cigarette consumption (yellow for at least one cigarette a day, green for nonsmokers); node sizes are proportional to the number of cigarettes smoked; and arrow colors indicate the kind of relationship (orange for friends and spouses, purple for family). The close-up shows that smokers are more likely to be located at the periphery of their networks.

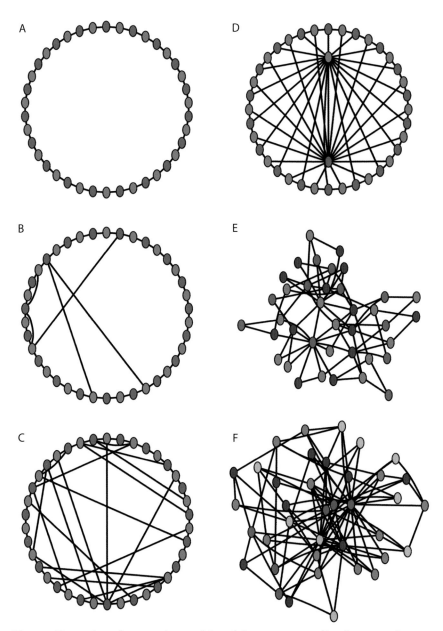

Plate 5. Examples of networks used in a laboratory coordination experiment. Each participant was assigned a particular location within one of these six pre-defined networks (all of which had exactly thirty-eight positions). Participants could see only their own color and the colors of those to whom they were directly connected. Participants were given a menu of colors to choose from and were allowed to change their color as often as they wanted. They were paid only if, within a fixed period of time, no two connected participants chose the same color. Here we show one solution to this "coloring problem" for each network. Different network structures (A–F) had important effects on the ability of the groups to coordinate a solution.

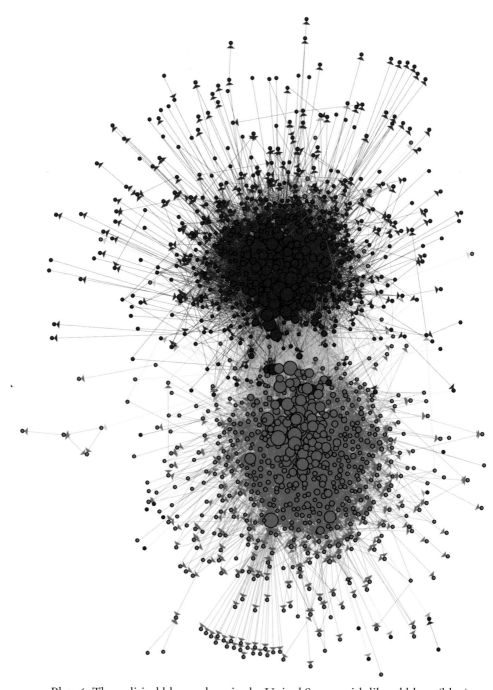

Plate 6. The political blogosphere in the United States, with liberal blogs (blue) and conservative blogs (red) and the web links between them. Tie color indicates the kind of link (blue between two liberal blogs, red between two conservative blogs, orange from liberal to conservative, and purple from conservative to liberal). Node size indicates the number of other blogs that link to it. This network map shows that the political blogosphere is highly polarized.

Plate 7. The Iranian political blogosphere. Distinct colors indicate communities of blogs that link much more to one another than to the rest of the blogosphere (labels indicate the name researchers gave each community). The two circled nodes are blogs written by two important political leaders, former President Mohammad Khatami and President Mahmoud Ahmadinejad. (Courtesy Morningside Analytics.)

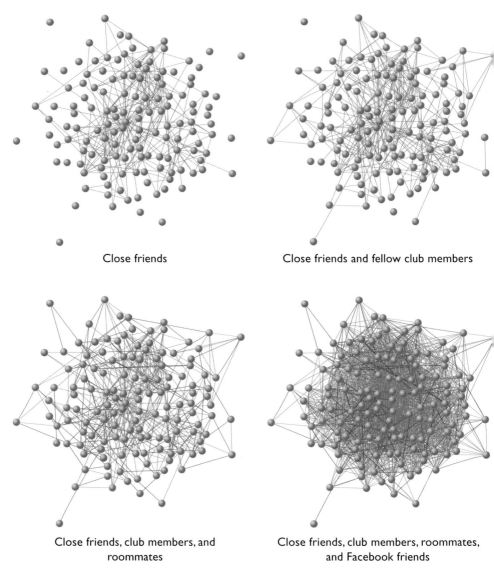

Close friends

Close friends and fellow club members

Close friends, club members, and roommates

Close friends, club members, roommates, and Facebook friends

Plate 8. An illustration of the difference in connectedness between real-life and online networks for 140 university students. The network at top left shows close, real-life friendships in gray. The top right adds the network of ties between people who belong to the same club in green. The bottom left adds the network of roommates in blue. Finally, the bottom right network includes all these relationships and adds the numerous online Facebook friendship relationships in orange. Online ties clearly greatly outnumber in-person connections and can even obscure real relationships. These images also illustrate the multiplexity of relationships in that some pairs of individuals might be friends, fellow club members, and roommates, and some not (in fact, it is a problem when two roommates are not friends).

is an exact tie. Otherwise, the outcome will not change whether you vote or not. So the value of voting is not $1,000; instead, it is a 1 in 10 million chance to obtain the $1,000 value.

Third, and finally, you have to compare your anticipated benefit to the costs of voting. Most people say that the costs of gathering information and going to the polls are not that great, so for convenience let's assume they are $1. They could be much higher, of course, but they are almost certainly greater than zero.

Hence, now that we have your costs and benefits all worked out, the rational analysis of voting suggests that the decision to vote equates roughly with the decision to pay $1 for a lottery ticket that gives you a 1 in 10 million chance of winning a $1,000 prize. Las Vegas would love to sell these tickets. If they could sell 10 million tickets, they would make $10 million dollars and owe just $1,000 in prize money. But even the most ardent gambler would probably refuse to buy them, knowing that the odds are extremely unfair. The average person would probably need other inducements to buy a ticket, because slot machines, blackjack tables, and roulette wheels all have vastly better odds. Even state lotteries that use funds from ticket sales to provide public services rather than prize money typically offer people millions of dollars in winnings, not thousands, for odds like these. And so we are left with the same puzzle we began with. Why do millions of people vote in spite of these odds and payoffs? What is it about elections that make them different from lotteries?

This rational analysis of voting is extraordinarily depressing for (at least) three reasons.

First, it suggests that the core act of modern democratic government makes absolutely no sense. Economists would call voting irrational because it violates the preferences of the people who engage in it. For some reason, people decide to vote even though they would not buy a lottery ticket with identical odds, cost, and payoff.

Economists typically think that people who vote are making a mistake or that there are other benefits to voting that we have not considered. For example, Downs himself noted that people might vote in order to fulfill a sense of civic duty or to preserve the right to vote. Later scholars have pointed out that people might vote because they enjoy expressing themselves—in the same way they enjoy expressing themselves when they cheer for their favorite team at a ballgame.

Second, learning about the irrationality of voting actually depresses turnout. In 1993, Canadian political scientists André Blais and Robert Young gave a ten-minute lecture on the rationality of voting to three of their classes and compared their students' voting behavior to that of students in seven other classes who did not hear the lecture. Perhaps not surprisingly, the students who heard the lecture were significantly less likely to vote.[7] Meanwhile, back in the United States, on Election Day in 1996, the *Lawrence Journal-World* published a guest column by Kansas University political scientist Paul Johnson about his reasons for not voting. He outlined the rational argument and noted that because of it he had not voted in the past thirty years. Within days there were several pointed letters to the editor denouncing his opinion and openly calling for his dismissal from the university. Johnson was not fired, but he did register to vote a week later in part to calm the controversy.[8]

Third, the inability to explain the decision to vote calls into question the rational analysis of all political behavior. Since we cannot use cost-benefit analysis to explain something as basic as voter turnout, some scholars argue that it makes no sense to apply rationality to other decisions such as who to vote for, whether to run for office, how to bargain with political adversaries, and so on. Instead of making rational choices that account for the costs and benefits of their actions, political actors might be affected by their emotions or by specific contexts that could not be generalized. In 1990, Stanford professor Morris Fiorina (one of William Riker's students from Rochester) dubbed this perplexing voting problem "the paradox that ate rational choice."[9] This is academics' way of saying that it makes no sense.

We Do Not Vote Alone

It was in this highly charged environment that we began our own work on voter turnout. We thought that scholars on both sides of the rationality debate were missing a crucial point: people do not decide in isolation whether or not they will vote. Thinking about the problem from the perspective of the individual voter misses the big picture entirely.

A large body of evidence suggests that a single decision to vote in fact increases the likelihood that others will vote. It is well known that when you decide to vote it also increases the chance that your friends, family, and coworkers will vote.[10] This happens in part because they imitate you (as discussed in previous chapters) and in part because you might make direct appeals to them. And we know that direct appeals work. If I knock on your door and ask you to head to the polls, there is an increased chance that you will. This simple, old-fashioned, person-to-person technique is still the primary tool used by the sprawling political machines in modern-day elections. Thus, we already have a lot of evidence to indicate that social connections may be the key to solving the voting puzzle.

Yet, even these insights about the social determinants of voting never went beyond the first step. Just like Anthony Downs and the other modelers who assumed all individuals were independent, the scholars who noted the social influence of one person on another assumed that pairs of people also acted independently. If I vote, it may help to influence my wife to vote, but the buck stops there. Scholars never wondered what might happen if they considered larger groups of people. Maybe the key to why we vote—and why it is rational for us to vote—is that we are all connected in larger networks.

As a child of the 1970s, James watched far too much television. He remembers one commercial in particular in which a woman is so enthusiastic about her new shampoo that she tells two of her friends about it. The screen then splits to show the two friends, and the voice-over says, "And she told two friends…and she told two friends…and so on…and so on…and so on…" The number of women on the screen doubled each time the voice-over said "she told two friends" so that by the end of the commercial sixty-four women were all using the shampoo.

This commercial is still used today to illustrate social marketing, but the idea that intrigued us was this: What if we replaced the act of trying out a new shampoo with the act of voting? What if a single act of voting not only influences my friend but also my friend's friend? One person might have just five friends, but if each of those friends also has five friends, then maybe a single person can influence all twenty-five of them as well, and all 125 of their friends' friends. It is easy to see how the number of people affected by a single decision can go up quickly. With an average of ten friends and family per person, we could easily imagine having influence over ten, then one hundred, and then one thousand people. And if one vote led to not just ten but to hundreds or thousands of votes, then maybe the likelihood of affecting the outcome of the election would be magnified enough to explain why so many people do vote. We may not see all the people we affect, but we may sense that our vote really might make a difference.

The earliest research on the social spread of political behavior came in the classic voting studies of Columbia social scientists Paul Lazarsfeld and Bernard Berelson that took place in the 1940s in the towns of Erie, Pennsylvania, and Elmira, New York.[11] Although they did not collect information about the whole network that interconnected all their subjects, they did ask people to discuss who influenced them and by what means, and this gave us the very first picture of how important networks can be in political behavior. One of the key findings from these studies was that the media does not reach the masses directly. Instead, a group of "opinion leaders" usually acts as an intermediary, filtering and interpreting the media for their friends and family who pay less attention to politics. In other words, the media appeared to work by getting its message to those who are most central in the social network. Politicians themselves follow a similar strategy, seeking endorsements from local leaders and targeting frequent voters rather than trying to persuade those at the periphery of the network who may or may not participate.

Later research by Robert Huckfeldt and John Sprague in the 1970s, 1980s, and 1990s would innovate on these earlier studies.[12] Their research in South Bend and Indianapolis, Indiana, and St. Louis, Missouri, used a "snowball" design, asking people to talk about friends who influenced them and to give the researchers their friends' contact information so they could be in the study too. Huckfeldt and Sprague found that when it came to politics, birds of a feather flocked together. Democrats tend to be friends with other Democrats, and Republicans tend to be friends with other Republicans. Liberals are connected to liberals, and conservatives to conservatives. Voters tend to discuss politics with others who vote. That is, people appear to be clustered together politically, acting and believing in concert with the people who surround them.

We wondered whether this insight could shed light on why people vote at all. We also wondered whether a strong similarity in people's local networks could arise from a spread of political behaviors and ideas. Did people choose to associate with those who resembled them, or did they induce a resemblance by influencing their peers? Huckfeldt and Sprague showed us the person-to-person effect, but now we wanted to know how and whether it might spread to other people in the network. Could one vote really spur thousands of others to vote in a turnout cascade?

Real Politics in a Social World

In order to find out just how far we could push the idea that voting might spread from person to person to person, we decided to try to answer the question, if I vote, how many other people are likely to vote as well? Many interactions between friends and family might affect the decision to vote. People might be affected by merely observing their acquaintances' behavior (Do they vote? Do they participate in community or group activities? Do they have political signs in their yards?). They might also be affected by political discussions with their acquaintances. Even chance encounters might be influential. As Huckfeldt writes: "The less intimate interactions that we have ignored—discussions over backyard fences, casual encounters while taking walks, or standing in line at the grocery store, and so on—may be politically influential even though they do not occur between intimate associates."[13]

Several election studies show that we typically talk to only a few people about politics; in one study in which people were asked to name their "discussion partners," about 70 percent reported fewer than five (on any topic).[14] Subjects reported talking with each of their discussion partners about three times per week, and most people said they talked about politics "sometimes" or "often." And while elections are not always on people's minds, a large number of people reported paying attention to campaigns, especially in the few months immediately prior to an election. Using data from a variety of sources, we estimated that respondents typically have about twenty discussions during the most salient period of a campaign when people are trying to decide whether to vote, but the number of opportunities for influence is probably even greater. A significant percentage of respondents in the Indianapolis/St. Louis Election Study—fully 34 percent—said they tried to convince someone to vote for their preferred candidate, indicating that many people believe there is a chance others will imitate them. These efforts might be aimed at influencing voter choice, but they also convey messages about whether an election is important, which might affect people's decision to show up on Election Day.

But do these attempts to influence others succeed? If imitation is occurring, then we should see a correlation in behavior between two people who are socially connected. And, in fact, that is exactly what we do see when it comes to voter turnout. Even when we control for alternative sources of similar behavior, such as having the same income, education, ideology, or level of political interest, the typical subject is about 15 percent more likely to vote if one of his discussion partners votes. But does this influence spread beyond that to the rest of the network? As it turns out, we see a correlation between people who are directly connected and also between people who are indirectly connected via a common friend. In other words, if you vote, then it increases the likelihood that your friends' friends vote as well.

Scholars who study voting behavior consistently find that people tend to segregate themselves into like-minded groups. As a result, most social ties are between people who share the same interests. When people with ideological or class-based interests are not surrounded by like-minded individuals in their physical neighborhoods and workplaces, they tend to withdraw and form relationships outside those environments. In the Indianapolis Election Study, about two out of every three friends had the same ideology as the respondent. In fact, we can even see this on a large scale in recent U.S. elections by looking at the increase in polarization between Republican "red states" and Democrat "blue states."

Ideological polarization does not affect total turnout, but it does affect how one vote can transform into many votes for your favorite candidate. If liberals and conservatives live side by side, well mixed throughout the population, then a turnout cascade has an equal chance of sending each kind of person to the polls. You might be conservative, but your liberal friend copies you by voting (maybe just to spite you!), and his liberal friend copies him, and her conservative friend copies her, so that by the time the cascade runs out of steam, your vote has influenced an equal number of liberals and conservatives to vote. The net effect would be two extra votes for the liberal candidate and two extra votes for the conservative candidate, so on balance there would be no change. With polarization, however, turnout cascades are more likely to affect like-minded individuals and yield extra votes for your preferred candidate. Suppose instead that your friend is conservative and so is his friend and so is her friend, so that your decision to vote generates four extra conservative votes and no extra liberal votes. If you knew you could get lots of people to support your favorite candidate just by voting, you would probably be more likely to do it than if you thought your vote would get canceled out by a mix of people from the Left and Right. This means that in ideologically polarized environments, the incentive to vote might be magnified by the number of like-minded individuals you could motivate to go to the polls.

Using everything we learned from Huckfeldt and Sprague about real networks of political interactions, we created a computer model to simulate what happens to the whole network when one person decides to vote.[15] In each simulation, we let everyone in the network try to influence the people to whom they were connected. We then measured the cascades in which one vote turned into two and then four and then eight, just like the shampoo commercial. Repeating the model millions of times allowed us to see the likelihood of turnout cascades and how many people a person could typically affect through her own behavior.

The results were very surprising. In some cases, one person's vote spread like wildfire, setting off a cascade of up to one hundred other people voting, even though people typically were directly connected to only three or four other people. On average, one decision to vote would motivate about three other people to go to the polls. Moreover, because liberals tend to associate with liberals and conservatives with conservatives, these cascades would yield sizable increases in the number of people voting the same way. Most of the time, one person's vote turned into two or more additional votes for their favorite candidate. So it seems that the more polarized we become by befriending only people with similar ideologies, the more motivated we are to participate in politics. This certainly creates a dilemma for people who think polarization is bad and voter turnout is good.

Interestingly, the total number of people voting had virtually no effect on how far the cascades would spread in our computer model. We originally believed that the size of turnout cascades would be bigger in larger populations because of the increased number of people who might be influenced. Instead, we discovered that turnout cascades are primarily local phenomena, occurring in small parts of the population within a few degrees of separation from each individual. As our Three Degrees of Influence Rule suggests, the power of one individual to influence many is limited by the effect of competing waves of influence that emanate from everyone else in the network.

Turnout in the Real World

People usually wonder whether computer models like this have meaning in the real world. No one has ever witnessed a turnout cascade, so how can we prove that they exist? Maybe they are just a figment of the modeler's imagination.

Many results from the model made sense and were already well established. The more often you ask someone to vote, the more likely it is she will vote. This is not surprising. What we needed was an important counterintuitive result that we could verify in the data. And, in fact, one prediction made by the computer model was very subtle and had never before been considered by political scientists. The model suggested that cascades would be largest if they emanated from someone who was in a moderately transitive group (that is, a group where people's friends know each other). Too much transitivity would mean the group was cut off from the rest of the world, and too little transitivity would mean the group was too disorganized to reinforce its own members' behavior. People might not know exactly how all their friends are connected, but they probably do have a sense of whether they might be able to reach people beyond their own group.

Hence, if there is a sweet spot for producing voting cascades, then we should expect people in that spot in real life to actually be more likely to vote because they are in a better position to influence many other people to follow suit. By the same reasoning, we should expect the same people to be more likely to try to persuade others to vote. As it turns out, this is exactly what we found in the Indianapolis and St. Louis data. The people who were the most likely to vote were those with transitivity of about 0.5 (meaning that half their friends are friends with one another). People whose friends did not know one another participated less, but so did those in tightly knit cliques of friends. And we recently confirmed these results and found exactly the same effects in a nationwide Gallup survey of networks and voting behavior.

These findings contradict some of the core recommendations made by political scientist Robert Putnam and his colleagues who study the effect of "social capital" on the health of our democracy.[16] Putnam argues that highly clustered network ties improve information flow and increase reciprocity at a societal level because everyone is looking out for everyone else. In other words, more tightly knit connections are better for society. However, our work shows that, at a certain point, networks can become so transitive that norms and information simply circulate within groups rather than traveling between them. Like Brian Uzzi's groups of scientists and Broadway musical producers that we discussed in chapter 5, democratic citizens work best in "small worlds" where some of our friends know one another and others do not.

While our computer model provided some of the first indirect evidence that turnout cascades are real, direct evidence was not far behind. In 2006, Notre Dame political scientist David Nickerson traveled to neighborhoods in Denver, Colorado, and Minneapolis, Minnesota, to conduct a novel experimental study of voter turnout.[17] In this study, experimenters walked door-to-door to contact people who lived in two-person households. Each of these households was randomly assigned to receive a "treatment" or a "control" message. In the treatment, the experimenter encouraged the person who answered the door to vote at an upcoming election. In the control, the experimenter encouraged the practice of recycling. Nickerson noted who came to the door to speak to the experimenter, and then waited until after the election to look up who had voted and who had not.

Voter-contact studies are very common, and it is well established that get-out-the-vote campaigns actually work. So it was not surprising that the people in Denver and Minneapolis who answered the door and heard the plea to vote were about 10 percent more likely to turn out than those who heard the plea to recycle. The big surprise, however, was in the behavior of the people who did *not* answer the door. As it turns out, the other person in the household was about 6 percent more likely to vote. In other words, 60 percent of the effect on the person who answered the door was passed on to the person who did not answer the door.

Consider for a moment how these indirect effects might flow through a whole network. Nickerson's creative study showed that a single plea to vote can change political behavior and spread from the experimenter to the person who heard the get-out-the-vote message to a person who neither heard the message nor met the experimenter. But why would it stop there? The person who didn't answer the door might pass the effect on to her other friends and family as well. The effect probably won't be as strong when it gets passed along; as in the game of telephone, the get-out-the-vote message might get diluted along the way as it passes from person to person to person. But suppose that the effect decreases the same way between every pair of people, going from a full effect to 60 percent at every step. If the first person is 10 percent more likely to vote and the second person is 6 percent more likely to vote, then the third person would be 3.6 percent more likely to vote, the fourth person would be 2.16 percent more likely to vote, and so on.

That may not seem like much change, but remember that while the size of the contagious effect decreases at each step, the number of people affected increases exponentially at each step. In a world where every friend has only two other friends, there might be only two people who become 10 percent more likely to vote, but there will be four more who are 6 percent more likely, 8 who are 3.6 percent more likely, sixteen who are 2.16 percent more likely, and so on. Add all those up for a city about the size of Denver or Minneapolis, and the result is that a single appeal to vote could cause about thirty extra people to go to the polls. And if you make an appeal to about three dozen people, suddenly you can get an extra thousand to go to the polls.

Of course, in real social networks, we tend to have more than two friends, and this increases the number of people who are close enough to us to be strongly affected by our actions. Yet, as we have noted, many of our friends already know one another, which means the effect might bounce around between the same people and never reach others who are socially distant from us. Also, the message may decay more rapidly than Nickerson found. It's hard to say which of these features of real-world social networks would dominate, but Nickerson's study gives us a taste of the potentially large cascade that could result from our personal decision to cast a vote.

Doing Your Civic Duty

So where do these results leave us on the question of why people vote? The existence of turnout cascades suggests that rational models of voting like those proposed by Anthony Downs, William Riker, and others have underestimated the benefit of voting. Instead of each of us having only one vote, we effectively have several and are therefore much more likely to have an influence on the outcome of an election. And the fact that one person can influence so many others may help to explain why some people have such strong feelings of civic duty. Establishing a norm of voting with our acquaintances is one way to influence them to go to the polls. People who do not assert such a duty miss the chance to influence people who share similar views, and this tends to lead to worse outcomes for their favorite candidates. In large electorates, the net impact on the result might be too marginal to create a dynamic that would favor people who assert a duty to vote. However, as Alexis de Tocqueville noted almost two hundred years ago, the civic duty to vote originated in much smaller political settings, such as town meetings, where changing the participation behavior of a few people would have made a big difference.[18] Actually, as we will see in chapter 7, social collaboration has an even more ancient origin.

And the norm of voting appears so deep-seated that many people are dishonest when they talk to pollsters. Typically, about 20 to 30 percent of the people who say they voted in an election actually did not. How do we know this? The ballot in America is secret, but whether or not you showed up to cast a ballot is a matter of public record, so we have excellent official information about who voted and who did not. The problem of overreporting voter turnout is very well known among political scientists and a common subject in college classrooms.[19] One of our favorite moments in Poli Sci 101 occurs when we ask our students to raise their hands if they did not vote. Typically, less than a quarter raise their hands. Yet, these are the honest ones, since we know from voter records that probably more than half the class did not vote.

Why do people lie about this? One possibility is that they fear social sanctions. Another is that they believe others are influenced by their political actions. Consider what happens if you tell everyone you are voting, but you stay home instead. On average, your actions will increase turnout even though you did not vote yourself. Moreover, since most of the people who decide to vote are likely to share your ideology, you can increase the vote margin for your favorite candidates without actually going to the polls. So now we have an explanation for why people might lie about voting as well. But, most important, we have an essential explanation for why people vote: they are connected, and it is rational for them to vote precisely because of their connections.

From Little Guys to Fat Cats

Voters are not the only political actors influenced by their social networks. If anything, the networks of politicians, lobbyists, activists, and bureaucrats are even more critical for determining how we govern ourselves. In fact, we want our political representatives to be well connected so they can influence others. And, indeed, politicians typically advertise their relationships with other important people. Every handshake is meticulously photographed, and many campaigns feature images of the candidate consorting with the powerful and the wealthy to make the point that this is a person who has the ability to get things done.

However, voters also worry that their representatives are connected to the wrong kinds of people. In the wake of the U.S. Congress influence-peddling scandal that erupted in late 2005, lobbyist Jack Abramoff was accused of buying votes and was widely described by the press as the "best-connected" lobbyist on Capitol Hill. President George W. Bush and other politicians like House Speaker Dennis Hastert and Senate Majority Leader Bill Frist feared the extent to which they could be "connected" to Abramoff, prompting them to return campaign donations and deny having spent time with him. They even curtailed contacts with other lobbyists to avoid the appearance that they were in some way influenced by lobbyists and legislators who had been tainted by the scandal.

Here we have a problem that we did not have with voters. Politicians know they are being watched, so they have an incentive to misrepresent their social networks. They might show a picture of a meeting with the president, but the president might have no idea who they are. They might hide a relationship with a powerful lobbyist or a steamy intern until they are forced to admit it under oath. And they might choose their friends (and even their spouses) in order to win elections. In other words, successful politicians tend to manipulate their networks for political advantage. This makes it nearly impossible to use the same methods to study politicians that we used to study voters. If we want to know who a voter's friends are, we just ask them; they have very little reason to lie. If we want to know a politician's connections, we need to use a little more ingenuity.

Following the Paper Trail

Although legislators do not broadcast lists of their enemies and friends, they do leave an enormous paper trail that we can study for clues. Some of the earliest attempts to discover relationships between politicians defined a connection as frequency of agreement on roll-call votes. The idea is that if Democrats Hillary Clinton and Barack Obama always vote "aye" on exactly the same bills, then maybe that means they are connected and that they might possibly be friends. However, agreement might also just mean these legislators have the same opinions about what laws should be passed. Clinton and Obama may both vote for a health-care bill they like, but they still might refuse to talk to each other. So roll-call votes might be more about ideology than personal relationships. Political scientists Keith Poole and Howard Rosenthal have developed highly sophisticated techniques that show that voting records can be used to place politicians on a liberal-conservative scale.[20] They find that the ideological divide between Democrats and Republicans is huge and growing, but this does not necessarily imply a friendship divide. If we relied on roll-call votes to try to discover the social network between senators and representatives, we would have missed countless cross-party connections that we know exist, like the close friendship between Democrat Patrick Leahy and now former Republican Arlen Specter.

Instead of roll calls, we decided to look at a different activity. Whenever a bill is introduced in the House or Senate, the person who introduces it is called the "sponsor." Legislators then have an opportunity to express support for it by signing on as a "cosponsor." Sponsors tend to spend a lot of time recruiting cosponsors by making appeals to other legislators in person and in "Dear Colleague" letters. They do this not only because it increases the chance the bill will be passed but also because it helps them win elections. They also frequently refer to the cosponsorships they have received in floor debate, public discussion, letters to constituents, and campaigns. For example, when then-Senator Barack Obama tried to persuade fellow members of the Senate to pass his bill on government transparency, he noted the bill had been "cosponsored by more than forty of our colleagues."[21]

The act of cosponsorship contains important information about the social network between legislators. In some cases, cosponsors actually help write or promote legislation, which is clearly a sign that the sponsor and cosponsor have spent time together and established a working relationship. In other cases, cosponsors merely sign on to legislation they support. Although it is possible that this can happen even when there is no personal connection between the sponsor and the cosponsor, it is likely that legislators make their decisions, at least in part, based on the personal relationships they have with the sponsoring legislators. The closer the relationship, the more likely it is that the sponsor has directly petitioned the cosponsor for support. It is also more likely that the cosponsor will trust the sponsor or owe the sponsor a favor, both of which increase the likelihood that the cosponsor will sign the bill. Thus, on average, cosponsorship patterns are a good way to measure the social connections between legislators.

Our cosponsorship network project was one of the first in political science to take advantage of the new era of large-scale data collection.[22] The Library of Congress regularly keeps tabs on bills in the Congress, so we had access to more than 280,000 pieces of legislation proposed in the U.S. House and Senate since 1973, and these bills involved roughly *84 million* cosponsorship decisions. There are many ways to use these data to measure how much total support a legislator receives from other members of Congress. The simplest is to count the total number of cosponsorships each legislator receives. If politicians are more influential, they should be better at getting their colleagues to support their bills.

Interestingly, the very first time we used this objective method for measuring influence, the legislator who turned out to be most influential was more corrupt than charismatic. The representative who received the most support during the 2003–2004 session of the House was Randy "Duke" Cunningham, a representative from Southern California who, according to the *Washington Post*, was involved in "the most brazen bribery conspiracy in modern congressional history."[23] Cunningham sold his house to defense contractor Mitchell Wade, who paid much more than the property was worth (Wade quickly resold the house at a loss of $700,000). Shortly thereafter, Wade began to receive defense contracts worth millions. Cunningham also lived on a yacht owned by Wade, and, according to the *Wall Street Journal*, he was provided with prostitutes, hotel rooms, and limousines in exchange for defense contracts. He eventually pleaded guilty to tax evasion, conspiracy to commit bribery, mail fraud, and wire fraud in federal court, where he was sentenced to one hundred months in prison (the longest sentence ever handed down to a former U.S. congressman).

Another interesting feature of the data was the degree of mutual support. If cosponsorships were really saying something about personal relationships, then we would expect to see a lot of reciprocity ("You scratch my back, and I'll scratch yours"). Here, we measured the number of times one legislator cosponsored another and then compared that to the number of times the sponsor returned the favor. Not surprisingly, the rate of mutual cosponsorship was quite high, especially in the "good ol' boys" network of the Senate, and the pattern has remained very consistent since the early 1970s.

Since cosponsorships indicate how well two people work together, we can also use them to say something about the network as a whole. Consider the observation that Americans have become increasingly polarized between Democrats and Republicans during the past few years. If that is really true, then, over time, we ought to see fewer cosponsor/sponsor relationships that cross party lines compared to the relationships that stay within party lines.

Imagine a network in which Democrats only work with other Democrats, and Republicans only work with other Republicans. The illustration shows Democrats and Republicans as two completely separate communities, or *modules*. Now suppose a few Democrats start crossing the aisle, or vice versa. This network would be less modular; it would be less obvious that there are two clearly defined groups that tend not to work together. In the extreme case where Democrats work with Republicans as often as they work with other Democrats and vice versa, the picture would look like one big network and it would not be possible to identify any modules at all. So the more modular the network is, the more polarized it is.

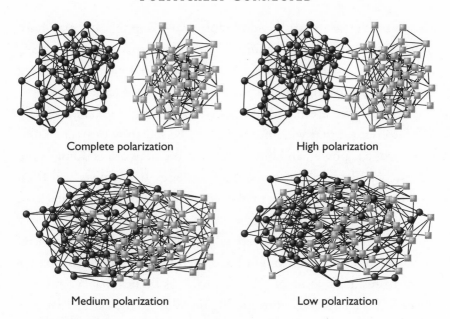

Complete polarization   High polarization

Medium polarization   Low polarization

Hypothetical networks of one hundred U.S. senators reveal how polarized they can be. Black circles indicate Democrats, white squares indicate Republicans, and the lines between them indicate collaborative relationships.

Physicist Mark Newman has developed some ingenious algorithms to measure modularity and find coherent communities in social networks, and in our own research we used one of these to see how polarization in the U.S. House and Senate has changed over time.[24] The results showed a sharp increase in the late 1980s and early 1990s, but then it leveled off. Some people blame House Speaker Newt Gingrich and the "Republican Revolution" for the dramatic rise in polarization. He and fellow Republicans Tom DeLay and Bill Frist swept to victory with their "Contract with America" in 1994, upset the seniority system in order to give Republican freshmen more power on committees, and worked to consolidate Republican districts in Texas and other states to maintain control of the House of Representatives. However, the network analysis shows that polarization started to increase sharply well before 1994. Republicans may have contributed to partisan breakdown, but evidence from the social network suggests that the change in leadership in 1994 was part of a broader trend toward a more polarized political system.

Although we are highly polarized in America, we are no more likely to remain that way than we are to be unhappy or overweight. Knowledge is power, and knowing what the network is doing is the first step toward solving potential problems. If scientists had been able to track polarization in the cosponsorship network back in 1990 and 1992, perhaps American citizens would have received early warning of the changes that were taking place in the legislative social network, and perhaps it might have been possible to make a better effort to head off some of the more vicious battles that would mar the political landscape in the coming decade—to the extent that network factors were responsible. For example, perhaps the Democratic leadership in 1992 would have worked harder to cross the aisle if they had known doing so would prevent the large-scale changes to the American political system that would lead to their twelve-year exile from power.

The Best-Connected Politician

Social-network theorists have described a variety of ways to use information about social ties to measure the relative importance of group members. But none of these measures takes advantage of one other piece of information available here: the strength of the relationship between legislators.

Counterintuitively, the best bills in Congress for measuring social relationships are the ones that receive the least support. Why? Because bills with many cosponsors (sometimes called "Mom and Apple Pie" bills by political scientists) are frequently supported by legislators who had no contact with the sponsor. For example, ninety-nine senators signed Ted Kennedy's bill "honoring the sacrifice of members of the United States Armed Forces who have been killed in Iraq and Afghanistan." In contrast, bills with just a few cosponsors signal that the sponsor and cosponsor worked together or knew each other well. For example, in 2003, Representative Edward Schrock from Virginia was the sole cosponsor on a bill sponsored by Todd Akin from Missouri. A brief visit to their personal websites revealed that Schrock and Akin worked together on the House Committee on Small Business, and each mentioned his collaborative relationship with the other.

We thus used signatures on bills with few cosponsors to infer connections between legislators and to draw the network of supporting relationships. When we examined this network, we found confirmation of the adage "where you stand depends on where you sit." People who are officially supposed to work together tend to be close, even if they are from different parties. The majority and minority leaders often had very strong relationships (like Republican Bill Frist and Democrat Tom Daschle), as did committee chairs and their cross-aisle counterparts (like Republican Bob Ney and Democrat John Larson). We also found strong relationships between people from the same state or from neighboring districts (like representatives Jim DeMint from South Carolina and Sue Myrick from North Carolina). But sometimes closeness to members of the other party can be an early warning signal of a party switch. Republican Senator Arlen Specter from Pennsylvania was so deeply embedded in relationships with Democrats in 2007 and 2008 that we originally thought we might have made a mistake in assigning his party when preparing the data. But, as it turns out, the network was just telling us that there was a good chance he would soon cross the aisle, which he did in early 2009.

We uncovered personal relationships too. For example, there was no official or geographic relationship between Senators John McCain and Phil Gramm, and their voting records differed on several important points. But the network of cosponsorships suggested they were very close in 2001 and 2002. And, as it turns out, McCain chaired Gramm's 1996 presidential campaign, and McCain has also publicly discussed his close friendship with Gramm, which started in 1982 when they served together in the House.[25] So there it was. The paper trail seemed to be leading to the very network we sought to discover.

Conversely, the network of cosponsorships also can be used to find personal enemies. Some representatives may have similar ideologies but nevertheless personal animosity—perhaps owing to a failed business dealing, a sexual indiscretion, or some other personal conflict. The feud between Democratic New Jersey Senators Frank Lautenberg and Bob Torricelli is legendary. At a closed-door caucus meeting of Senate Democrats held in 1999, Lautenberg chastised Torricelli for telling a reporter that he felt closer to Christie Todd Whitman (the Republican New Jersey governor) than his fellow Democratic senator. Torricelli became so enraged that he stood up and indecorously screamed, "You're a fucking piece of shit, and I'm going to cut your balls off!"[26] Not surprisingly, Torricelli and Lautenberg very rarely cosponsored each other's bills, in spite of their close ideological and geographic affiliations.

The network of cosponsorships allows us to see how well connected each legislator is to the other legislators in the network. The legislators at the center of this network read like a who's who of American politics, including Tom DeLay, Bob Dole, Jesse Helms, John Kerry, and Ted Kennedy. Without any specific information about these legislators other than their bill cosponsorship, the network reveals which individuals are most influential and which are more likely to run for higher office (our most recent top twenty list included Hillary Clinton, Ron Paul, Tom Tancredo, and Dennis Kucinich). And when we look at all the data, the very highest scoring legislator is John McCain, who was the 2008 Republican nominee for the presidency.

But the point of all this ranking and naming of names is not merely to arbitrate a contest about which cat is fattest. The reason we built the network and looked closely at the people at the center was to test the validity of our argument that the structure of a network matters. On its face, the network seems to identify party leaders, committee chairs, and other well-connected people. Legislators who are able to elicit support in the cosponsorship network because they are broadly connected or well connected to other important legislators ought to be better able to shape the policies that emerge from their chamber. And in fact, they are.[27] In the House, members in the center of the cosponsorship network passed three times as many amendments as those on the periphery. In the Senate, the difference was even greater; highly connected senators passed seven times as many amendments.

Being well connected in the social network makes a huge difference when it comes to shaping bills as they move through the legislative process. However, this tells us nothing about the success of those bills. Senators and members of the House can add all the amendments they want, but if the bill fails final passage, all is for naught. To what extent does connectedness influence the outcome of final votes on the floor? If better-connected legislators are indeed more influential, then they should be able to recruit more votes for the bills they sponsor. Otherwise, what is the point of being well connected in the first place?

When we looked at the effect of social-network position on roll-call votes, we found that the best-connected representatives were able to garner ten more votes than average (out of 435 representatives), while the best-connected senators were able to garner sixteen more (out of 100 senators). That may not seem like much, but consider how close many of these roll-call votes are. Changing the connectedness of the sponsor of a bill from average to very high would change the final passage outcome in 16 percent of the House votes and 20 percent of the Senate votes. In other words, if a bill is introduced by a person in the middle of the network, it would pass; but if the same bill were introduced by someone just outside of the middle, it would fail. Connectedness matters.

The Network Architecture of Political Influence

In addition to voters and politicians, lobbyists and activists also live in social networks that have a big impact on their effectiveness. It is well known that lobbyists tend to spend time with legislators with similar policy preferences, making us wonder what exactly it is that lobbyists accomplish. After all, a Haliburton lobbyist is not going to change Dick Cheney's mind any more than a Sierra Club representative will change Al Gore's. That's just preaching to the choir. The popular image of lobbyists is that they are engaged in influence, but instead it seems like they spend more time engaging in homophily, flocking together with birds of a feather.

Political scientists Dan Carpenter, Kevin Esterling, and David Lazer carefully studied the social networks of energy and health care lobbyists and found a much more nuanced story.[28] While it is true that lobbyists tend to develop strong ties with their ideologically similar counterparts in the government, their success is influenced by the network as a whole. For example, lobbyists are much more likely to communicate with one another if their relationship is brokered by a third party. Also, lobbyists are more likely to be granted access to key players in the government if they are connected to someone who already has access. So the greater the number of friends they have who already have access, the better. What this means is that the most successful lobbyists will be those who have the most weak ties, that is, the most friends of friends walking the halls of power. Strong ties help, but weak ties help more because they greatly expand the total number of potential connections. Just like job seekers tapping their weak ties (in chapter 5), searching for influence is easier with a broad network. In fact, Carpenter and his colleagues found that the number of strong ties has almost no impact on whether a lobbyist will be given access. Since each new weak tie can lead to multiple others, this sets up a rich-get-richer dynamic that generates rising stars like Jack Abramoff and helps to explain why corruption is so widespread.

While lobbyists are firmly embedded within the political system, the same is not always true for activists. Abbie Hoffman—member of the "Chicago Seven" and cofounder of a 1960s activist group called the Youth International Party (the "yippies")—encouraged his followers to work against the system and showed them how to grow marijuana, steal credit cards, and make pipe bombs.[29] American social movements are often bitterly divided over the question of whether to work for change within the system or outside it. Political scientists Michael Heaney and Fabio Rojas were interested in finding out why some worked within and others worked without, and not surprisingly they found that social networks played a critical role.

As the movement against the Iraq War was heating up in 2004 to 2005, Heaney and Rojas collected information from 2,529 activists at several events, including a 500,000-person protest outside the Republican National Convention in New York City on August 29, 2004; a protest of George W. Bush's second inauguration in Washington, DC, on January 20, 2005; antiwar rallies in New York City, Washington, DC, Fayetteville, NC, Indianapolis, Chicago, San Diego, and San Francisco, commemorating the second anniversary of the Iraq War on March 19 and March 20, 2005; May Day rallies held in New York City on May 1, 2005; and the 300,000-person antiwar protest in Washington, DC, on September 24, 2005.[30] Each of the activists provided basic information about why they were protesting and named organizations that had contacted them to attend the rally. This gave the researchers an incredibly detailed picture of the overall network of interactions and allowed them to make two important conclusions.

First, whether they admit it or not, activists' behavior is shaped by differing partisan attitudes, because they tend to join organizations with others who share their party affiliation. The "party in the street" may think of itself as quite disconnected from the type of formal party that runs the government, but it ends up attracting people who all have the same partisan ideologies. Second, and not surprisingly, activists who are more central to the network of political groups are more likely to work within the system, embracing institutional tactics like lobbying instead of civil disobedience. So people who think of themselves as Democrats might join the Sierra Club, but they are very unlikely to join less established groups such as the yippies that might be pursuing the same goal with different methods.

Activism Goes Online

When we first published the results of our voting model, a number of online activists became very interested in the idea that voting is contagious. In particular, GROWdems.com contacted us early on to include our research in an e-book they created to improve their get-out-the-vote efforts. They believed that knowing that one extra vote leads to many others would give volunteers a greater sense of purpose and effectiveness, which could motivate more people to help with their campaign. An online group at CircleVoting.com also started using our research to encourage people to reach out to their online social networks to get them to the polls.

But these efforts are just the tip of the iceberg. The Obama campaign's use of Internet and mobile technology shows the real power of online social networks. They took advantage of social media like YouTube for free advertising. Internet users watched a stunning 14.5 million hours of official campaign ads online. For comparison, it would have cost about $47 million to buy that much advertising time on broadcast TV. They also used YouTube to combat negative stories. When Obama's former pastor, Reverend Jeremiah Wright, made the news with his "God Damn America" sermon, the traditional media latched on to the story and covered it for several days. Meanwhile, supporters forwarded links to Obama's own speech on race, which made it hard to believe that he shared Wright's views. During the primaries alone, 6.7 million people watched Obama's thirty-seven-minute speech on YouTube.

Other candidates also tried to organize their supporters online, but with less success. The Pew Research Center reported that Obama supporters were more likely than Clinton supporters to watch campaign speeches and announcements, campaign commercials, interviews with the candidates, and televised debates online. They were also more likely to donate money online.[31]

Activists around the world are also starting to use the Internet to organize vast demonstrations. For example, in January 2008, Oscar Morales, a thirty-three-year-old engineer from Barranquilla on Colombia's Caribbean coast, mobilized millions of people by using his social network. On the social networking site Facebook, he started a group composed of himself and five friends (Hector, Juan, Miguel, Maritza, and Gabo) protesting the holding of hostages by the military group FARC (Fuerzas Armadas Revolucionarias de Colombia). Morales's group, called "No More," grew to include 272,578 online members within a month. Invitations to real-life marches spread through cyberspace, and the intensity built over weeks. Finally, on February 4, 2008, as planned, millions of people in countries around the world marched to protest the taking and keeping of hostages: 4.8 million people turned out in nearly four hundred events in Colombia, as well as hundreds of thousands of others in countries ranging from neighboring Venezuela to faraway Sweden, Spain, Mexico, Argentina, France, and the United States.[32]

The Colombia demonstrations illustrate the power of online social networks to magnify whatever they are seeded with. One person started a campaign that touched millions. But online activism started long before the Obama campaign and the advent of Facebook. In the early days of the Internet, people like Glen Barry used the new technology to write about and promote political causes. Barry's "Gaia's Forest Conservation Archives" was an online diary that commented on current events related to the environment and urged the government to preserve forests as early as 1993 (it can still be found online). Soon afterward, a variety of people were promoting many different causes in their web logs, or blogs, and the blogosphere was born.

Because information was so easy to transmit on the Internet, some people believed that the blogosphere might bring us closer together politically. The hope was that we would discuss issues of the day in a nearly ideal Jeffersonian form of democratic exchange. But Lada Adamic, a physicist at the University of Michigan, has produced some stunning images of these exchanges that show nothing could be further from the truth.[33] In plate 6 we reproduce her image, from the 2004 election, of the network of A-list bloggers from the Left and Right, including well-known sites like Daily Kos, Andrew Sullivan, Instapundit, and RealClearPolitics. Conservative blogs are red, as are links between them, whereas liberal blogs and links are blue.

What immediately stands out is the extreme separation between liberals and conservatives. If the hope for the Internet was that these two groups would talk to each other, the blog network reveals that these hopes have been utterly dashed. Just like the real-world political networks studied by Lazarsfeld and Berelson and later by Huckfeldt and Sprague, the online social network appears to be strongly homophilous and polarized. This suggests that political information is used more to reinforce preexisting opinions than to exchange differing points of view. Adamic used a procedure to detect "communities" in the network (similar to Newman's modularity procedure that we used for our cosponsorship work). Communities were defined as a group of blogs that were more closely connected to one another than to the rest of the network. She found that the conservative bloggers were much more densely connected to one another within their "community" than the liberal bloggers were, suggesting that the reinforcement effect is even stronger on the Right than on the Left.

But even though liberals may tend to seek out opposing points of view more often, the dramatic separation shows that, like conservatives, they still stick very closely to the ideas and facts they know.

In this example, there may not be much communication between people who support and oppose the government, but at least an opposition party is allowed to exist. Researchers at the Berkman Center for Internet and Society at Harvard Law School have extended Adamic's research to other countries in order to see whether their blogospheres follow similar patterns. The first country examined was Iran, where they collected daily information for seven months from nearly one hundred thousand Persian language blogs.[34] As part of the Internet and Democracy project, John Kelly and Bruce Etling were very interested in whether the blogosphere had a personal impact on freedom of speech or a more global impact on Iran's prospects for liberalizing their form of government. Given Iran's reputation for political repression, they expected to find a tightly controlled and repressive political discourse. But instead they found a network of blogs that was not much different from those found in the free world.

Plate 7 shows a map of the Iranian blogosphere. There are so many links that they are suppressed here to make it is easier to see what is going on. Once again, the larger nodes are the more important blogs (as measured by the number of links to the blog), but in contrast to plate 6 (the U.S. blogosphere map that showed only political blogs), here there are several "communities" that focus both on political and nonpolitical topics. As Kelly and Etling note: "Iranian bloggers include members of Hezbollah, teenagers in Tehran, retirees in Los Angeles, religious students in Qom, dissident journalists who left Iran a few years ago, exiles who left thirty years ago, current members of the Majlis (parliament), reformist politicians, a multitude of poets, and quite famously the president of Iran, among many others."[35]

The Iranian blogosphere divides into four more or less coherent communities that can be described by the content of their blogs. Two of these have little to do with politics or public affairs. One focuses mainly on poetry and Persian literature, and the other is a hodge-podge of special interests and popular topics (related to celebrities, sports, and minority cultures). But the other two groups are explicitly political.

The first political group is composed of two closely overlapping communities: a reformist politics community of internal dissidents and a secular/expatriate community that is made up of notable dissidents and journalists who left Iran and are living abroad. They tend to discuss women's rights, political prisoners, and current affairs, including political problems in Iran such as drug abuse and environmental degradation. Since much of this discussion is critical of the government, it is somewhat surprising that most bloggers use their own names rather than pseudonyms. A second political group, made up of conservatives and religious youth, blogs about their support of the Iranian Revolution and the government's Islamist political philosophy. One notable community in this group is the Twelvers sect that believes that Muhammad ibn Hasan ibn Ali (the 12th Imam) will return to save mankind and create a perfect society before a final day of resurrection. But this is not simply a bunch of yes-men: many of the conservatives in this group actually attack the government for being too corrupt or too lenient.

Notice that two of Iran's presidents have popular blogs. President Ahmadinejad is part of the conservative group, and his predecessor President Khatami is part of the reformist group, but both blogs are positioned near the center of the blogosphere since they tend to get referenced by blogs from a variety of communities. Their positions indicate that they sit on a number of paths from one community to another, acting like the "weak tie" bridges that characterize successful lobbyists and politicians in the United States.

In fact, the Iranian blogosphere looks quite a bit like the U.S. blogosphere, which is puzzling. How could an illiberal regime permit such a wide range of seemingly democratic discourse? The Iranian government does block access to several websites, but less than 20 percent of the reformist blogs are affected, and hardly any of the conservative blogs are affected, even those that are highly critical. This suggests that the government either cannot or will not shut down the discourse. It is hard to believe that the regime lacks will, given its record of shutting down traditional media sources (such as opposition newspapers) and jailing (or worse) the people who run them. But if so, then maybe the ability to relocate democratic social networks and the resulting flow of information to an online environment can thwart the government's attempts to disrupt these networks, control information, and prevent the self-organization of a political opposition. Indeed, in June 2009, the media was proclaiming a "Twitter revolution" in Iran as citizens used the Twitter microblogging service to disseminate information across online network ties and to organize against what seemed like a rigged election. But only time will tell if the Iranian blogosphere will have a liberalizing effect on the government there.

This suggests that changes in technology may be altering the way we live in our social networks and may have profound effects on the way we govern ourselves. We have already seen that real-world social networks can be used to spread information and enhance the ability of well-connected people to achieve their goals. In the next two chapters, we will look more closely at the nature and origin of our desire to connect and how technology may be changing the way we do so. In a certain sense, we live in a brand-new world. Our social networks are ever faster and larger as we text, e-mail, Twitter, Facebook, and MySpace all the people we know (and even people we don't). And this new world certainly gives us a bird's-eye view of the social networks in which we live, making us more conscious than ever of the importance of being connected. But it also seems to us that these networks were ready-made to be put online. We have lived in them for millions of years. Our ancestors prepared us to live in them. Networks are under our skin. And before we think about where we are going, it will be useful to pause and reflect on where we have been.

CHAPTER 7

It's in Our Nature

S ocial networks can be difficult to understand in part because they are difficult to manipulate. We cannot give you a friend the way we might give you a placebo. But if we could somehow strand a group of strangers on a desert island and watch how they become connected, and for what purpose, we might be able to observe social networks as if we were conducting an experiment. This does not sound like something that could be done. Except that it has been done, and not by curious social scientists but by television producers.

In the summer of 2000, CBS piloted *Survivor,* a show that would become a sensation and spawn the reality-television craze. The premise was simple: take sixteen average Americans, chosen from all walks of life, and abandon them in the middle of nowhere to fend for themselves. Every three days, the "Survivors" would gather in a tribal council to vote someone out of the tribe. At the end, a sole Survivor would win $1 million.

The highest-rated season of *Survivor* took place in the Australian outback in 2001, drawing nearly thirty million regular viewers who tuned in each week to watch the social dynamics, among other things, unfold. In the first episode, contestants were forced to remain silent until they arrived on site and were placed into two competing tribes, Kucha (an Aboriginal word for "kangaroo") and Ogakor (an Aboriginal word for "crocodile"). They had five minutes to gather their gear and start a grueling five-mile trek to their camps. Debb Eaton, a forty-five-year-old female corrections officer at a men's prison in New Hampshire, quickly took charge of the Kucha tribe, but her initiative would be her undoing. Her fellow tribe members became so annoyed with her bossiness that she was the first person voted out.[1]

Some Survivors quickly made friends, while others plotted to eject fellow tribe members. For example, Jerri Manthey, an actress from Los Angeles, accused Kel Gleason, an U.S. Army Intelligence Officer from Fort Hood, Texas, of smuggling in beef jerky and refusing to share it. Tribe members searched his belongings and found nothing, but because they were already suspicious of his background and trustworthiness, the damage had been done. Kel was unanimously voted out at the next tribal council.

Every three days, the Ogakor and Kucha tribes competed in "immunity challenges" that forced the losing tribe to eliminate one of its members. As a result, the strongest members of each tribe were highly valued because they could help other tribe members avoid early ejection. On the other hand, the strongest members were also targets for ejection since they would be harder to beat in head-to-head competition when the number of tribe members eventually dwindled. As players schemed to form coalitions, they lobbied their fellow castaways using arguments such as these to eliminate the strong and the weak.

During the third week, the two tribes merged. Under this entirely new social arrangement, immunity challenges pitted players against one another, and the winner could not be ejected at the next tribal council meeting. Now the goal was clear: get rid of the strongest individual player. We might think that this would automatically lead to the demise of the second strongest person (the one who did not win immunity), but another potent force was at play. Over the course of the preceding three weeks of struggling against the elements, each tribe had bonded, and these coalitions endured even after the merger.

Nowhere was the power of social connection more evident than at the first immunity challenge of the merged tribe. Contestants were forced to stand on wooden pillars above a river. The object was simple: the last person standing on a pillar would win immunity. One by one, the contestants gave up, jumped into the river, and swam ashore. Some gave up because they were exhausted. Others gave up when Jeff Probst, the show's host, offered to give them peanut butter in exchange. After nine uninterrupted hours of standing, there were only three contestants left, and then Alicia Calaway, a personal trainer from New York City, jumped in the water. The two remaining Survivors were Keith Famie, a professional chef from Detroit, Michigan, and Tina Wesson, a nurse from Knoxville, Tennessee. As it turns out, the two of them were both from the Ogakor tribe. But Keith was in greater danger of being voted out of the competition, and so he said to Tina, "I need this one," and she willingly dropped into the water. This gave him immunity and prevented the former members of the Kucha team from being able to coordinate their votes to cast him out. Tina later said, "It was harder to come in than to stand out there. I knew that for the good of our team, I had to let Keith win."

The season climaxed on day forty-one, when the remaining three contestants were Tina, Keith, and twenty-six-year-old Texan Colby Donaldson. Colby had just won immunity and would decide which of the other two would be eliminated at the tribal council. Tina was popular with contestants who had been previously ejected from the tribe—the very people who would decide which of the final two contestants would win. Meanwhile, Keith had alienated several players by insisting that his skills as a chef were necessary to their survival. Most viewers thought Colby would eliminate Tina so that he could compete against Keith. But he shocked the nation and probably gave up a million dollars when he chose to eliminate Keith instead.

After the vote that night, the episode concluded with many scenes from previous days that made it clear that Colby and Tina had done well in part because they had formed a very strong alliance that had helped them through many difficult situations. Colby had chosen friendship over what seemed like certain victory. And, indeed, when the final vote was tallied before a live studio audience in Los Angeles, Tina Wesson was declared the winner.

Many people questioned Colby's decision and claimed that he had miscalculated. But another plausible interpretation is that friendship and loyalty had trumped self-interest. This is exactly the dilemma that most of us face every day: Do we help our friends or help ourselves? And what are the consequences? Will we look dumb if we help others? Will we look mean if we do not? Is it possible to be nice and survive? And how can we possibly make these decisions when we have many friends in a dancing pattern of shifting alliances and interests?

Part of the fascination with the show was not only the characters but also their complicated interactions. *Survivor* presents a series of interlocking, connected biographies—a sociography, actually, something akin to a novel. And, like a Russian novel, the story follows the shifting connections between people in the group and all the social complexity, as well as the group's fluid morality. Shows like *Survivor* are alluring precisely because they mirror the ancient struggles within our minds and among our peers.

The Ancient Ties That Bind

Like ants, bees, penguins, wolves, dolphins, and chimpanzees, human beings are social animals, living in close proximity to one another in groups. In fact, the most important feature of the human environment is the presence of other members of our species. Because we lack any real predators, the only major threat to humans is other humans. If we did not need other humans so much, it would make a lot of sense to avoid them.

Our particular relations with other humans are therefore crucial. We deliberately choose to form social connections with specific individuals, with whom we share greater or lesser intimacy and affection, for brief or lengthy periods of time. And unlike other social species, we have a special capacity to imagine what others are thinking and feeling, including what they are thinking and feeling about us. Our embeddedness in social networks means that we must cooperate with others, judge their intentions, and influence or be influenced by them.

In short, humans don't just live in groups, we live in networks. In previous chapters, we have shown how social networks affect everything from emotions to health to politics. But the question remains: why do humans form these networks in the first place? Through several strands of new research, we are finding a surprising answer: our desire to form connections depends partly on our genes.

Evolution likely plays a role in the way we connect to one another because the very act of connection is itself subject to natural selection. Recall the burning house we discussed in chapter 1. You want to get water from the river as quickly as possible to douse the flames, and a group of people has to get organized to help you. Some network formations will work better than others. We might say that the one that works best is the most "fit" for the task.

Now suppose that we had a tournament where we let different groups (like Kucha and Ogakor) compete to put out fires. In each round, we'd set two fires, then let each team try to put one out as fast as it can. After each game, we would eliminate the team with the slowest response time and invite a new team to play. After many rounds of this, teams that play quite well would emerge, suggesting that they had figured out efficient ways to organize themselves and work together. The teams that ran randomly back and forth to the river would be eliminated, but the teams that coalesced in a bucket brigade would not. The envious and selfish teams would be eliminated, but the helpful and collaborative teams would not. In this way, the teams with the most fit social networks would become the winners. The structure of the social network would adapt to meet the functional requirements it needed to fulfill. We form a line because water needs to flow. And in fact, some species exhibit bucket-brigade behavior because it is an efficient adaptation to their environment; consider ants that pass food items from one worker to another, for example.[2]

In the modern world, we see tournaments that influence network structures all the time. In American football, the offense has to put seven of its eleven players on the line of scrimmage (just behind the ball) at the beginning of a play. However, the other four players can stand wherever they want. This freedom has led to an enormous variety of formations, including the T, I, ace, pro-set, single-wing, double-wing, shotgun, pistol, eastern shotgun, wishbone, flexbone, wing-T, and A-11. Some formations have specialized purposes, like the self-explanatory goal-line formation. Each has advantages, depending on which players need to interact and what the objective is. For example, the shotgun formation puts the quarterback farther away from the ball so that he has more time to pass to a receiver while avoiding rushers from the defense.

One reason for the enormous variety of offensive formations is that coaches constantly retool their strategies in order to beat commonly used defensive plays. Occasionally such innovation yields a completely new way of playing the game that the defense must then adapt to. Emory Bellard, an offensive line coach for the Texas Longhorns, invented the wishbone formation in the summer of 1968. Because the team had a quarterback, halfback, and fullback who could all run well with the ball, Bellard wanted to give the quarterback three choices (the "triple option") that could be executed after a play began. The quarterback would first decide whether to give the ball to the fullback who was close behind him. If he kept the ball, he would break in one direction. As the quarterback ran, he could cut inside or flick the ball to a halfback running with him in the same direction. With two halfbacks behind and flanking the fullback, the four men in the backfield made a pattern that looked like a wishbone, which led Mickey Herskowitz of the *Houston Chronicle* (and everyone else thereafter) to call it the wishbone formation. The extra halfback served as a blocker for the play, and the symmetry of the formation meant that the play could break left or right, so that the defense had to prepare for four different backs who might carry the ball.

Texas tied the first game it played with the formation and lost the next. However, Texas went on to win its next thirty games, including two national championships. Bellard took the successful strategy with him when he moved to Texas A&M University and then to Mississippi State University, and other teams later imitated the formation, including Oklahoma, Army, and Air Force. However, defensive coaches soon developed strategies to beat the wishbone, such as the backbone defense. These new counterstrategies eliminated the original advantage of the wishbone and forced offensive coaches like Bellard to continue searching for the perfect offensive line.

Naturally occurring social networks are not like football teams or reality shows. No coach tells us how to pick our friends, and no one eliminates us if we lose a game or fail to put out a fire. However, human beings are and always have been subject to an analogous set of constraints that determine what kinds of social-network structures work and endure. Among early hominids, individuals who lived in social networks that enabled a group to acquire more food or to fend off attackers were more likely to survive and reproduce. As a result, over a long period of time, the individuals who naturally formed networks or who had specific traits conducive to forming particular kinds of networks would have had a selective advantage and might eventually have made up the largest part of the population. The networks we form today exploit different tools (like cell phones and the Internet) and operate in a different environment, but the urges we have to connect and to organize ourselves into groups of friends evolved at a time when genetic evolution favored some patterns over others.

The Surprising Role of Connection in Cooperation

Natural selection is brutal. Only the "fittest" individuals survive and get to reproduce and pass their genes on to the next generation. This creates a puzzle when it comes to social interactions. Suppose you tend to share food with friends who are not related to you. If food is scarce, then your generosity will help your friends become more fit. And it will make you less fit (less able to survive). Therefore, the genes that contribute to your desire to share food are less likely to be passed on than are the genes that would make you keep all the food for yourself. So, where does sharing come from?

This is the puzzle of cooperation and altruism: people who are willing to help others should be, it might seem, less likely to survive than people who care only about themselves. More formally, *cooperators* who are willing to pay a personal cost to help a group of people are less likely to survive than *free riders* who do not pay a personal cost but benefit from the group's activities. For example, when humans learned to hunt large game hundreds of thousands of years ago, this gave a fitness advantage to the groups that knew how to do it. But if it is risky to take down a mastodon, why not let someone else do it? If you are the most selfish person in your group, then presumably you would be more likely to survive.

In one of the most celebrated openings to a novel in modern fiction, Ian McEwan, in *Enduring Love,* provides a vivid illustration of the problem of cooperation. A helium balloon is hovering near the ground in a green English field in strong winds. Curled up inside the basket is a frightened boy, and outside, hanging to a rope, is his grandfather, desperately trying to control the balloon before it is blown away. He calls for help, and five men come running. Each of the six men now has hold of a rope attached to the balloon, and no one is heeding the suggestions of any of the others; there is no leader to guide them. A new gust of wind comes, lifting the balloon ten feet off the ground, dragging the men into the air, dangling. If they all hold on, they will return to the ground swiftly and safely. But one lets go, and the balloon, freed of this ballast, jolts higher. Instantly, the others face a difficult decision. Another one lets go, then another, increasing the pressure on the others to follow suit, until just one remains. That man, a really good man, one in whom "the flame of altruism...burned a little stronger," holds on. He is carried away trailing the balloon like the tail of a kite until, to the horror and humiliation of the other men watching, he has no choice but to let go, three hundred feet above the ground, and fall to his death. As McEwan noted, these six men, hanging a few feet above the ground, "enacted morality's ancient, irresolvable dilemma: us, or me."[3]

The good news is that people very often ignore their selfish tendencies when interacting with people to whom they are connected. In *Survivor,* Tina could have forced Keith to continue standing on the pillar, but instead she cooperated and jumped in the water when he asked. And across a wide variety of laboratory experiments that study dilemmas in altruism and cooperation, people choose to help other people about half the time, even people they will never interact with again. So it seems that a naive application of evolutionary theory to whether it makes sense to help other people would be incorrect. Selfishness does not always pay. If it did, we would all be selfish.

Now, in the real world, outside the laboratory, there are many other considerations, because we live in a web of interactions with one another and because we have repeated or sustained interactions across time. Tina might have given up because she knew she would be competing with Keith in future challenges and would need his help. Evolutionary theorists call this *direct reciprocity*. If you have several opportunities to cooperate with the same person, one way to get that person to help you is to promise future cooperation.

In an ingenious and famous study designed to examine reciprocity, political scientist Robert Axelrod showed that a cooperative strategy called "tit for tat" often is more effective than always cooperating or always being selfish.[4] In tit for tat, you cooperate the first time you meet someone, and thereafter simply copy what that person did the last time you interacted with him. This behavior is basically an inverse Golden Rule: do unto others as they have done unto you. If someone cooperates, then reciprocate that cooperation the next time around. If someone does not cooperate, punish him next time by withholding cooperation. Simple, but effective.

In a world full of people using the tit-for-tat strategy, cooperation will happen all the time. But in a world of purely selfish people, a person engaging in tit for tat does not fare so well. If you are using the tit-for-tat rule, the first time you meet a selfish person you will cooperate with him but he won't cooperate with you. You learned your lesson, and you will copy him in any future interaction, but that first meeting means he is a little more likely to do better since he got something from you on the very first interaction. If you don't meet some other people who are willing to take a chance on cooperation pretty soon, you and your genes are doomed.

It was this situation that led mathematician Chris Hauert and his colleagues to consider another possibility in an important evolutionary model published in *Science* in 2002.[5] In Axelrod's study and in most previous theoretical models, individuals were forced to interact with each other. But what if they could choose not to interact? Rather than attempting to cooperate and risking being taken advantage of, a person could fend for herself. In other words, she could sever her connections to others in the network. Hauert called the people who adopt this strategy "loners."

Using some beautiful mathematics, Hauert and his colleagues showed that in a world full of loners it is easy for cooperation to evolve because there are no people to take advantage of the cooperators that appear. The loners fend for themselves, and the cooperators form networks with other cooperators. Soon, the cooperators take over the population because they always do better together than the loners. But once the world is full of cooperators, it is very easy for free riders to evolve and enjoy the fruits of cooperation without contributing (like parasites). As the free riders become the dominant type in the population, there is no one left for them to take advantage of; then, the loners once again take over—because they want nothing to do, as it were, with those bastards. In short, cooperation can emerge because we can do more together than we can apart. But because of the free-rider problem, cooperation is not guaranteed to succeed.

To deal with free riders, another type of person is needed: *punishers.* People everywhere feel the desire to enforce social norms they see being violated. Some people honk when a car cuts them off in traffic, even though the honk does not change the outcome. Others risk confrontation by asking people smoking in a no-smoking area to stop. And on many occasions, innocent bystanders are willing to testify in court to crimes they have witnessed, even though this potentially exposes them to retribution. These people all pay a small cost themselves to impose a cost on someone who does not cooperate. And this is a different kind of connection. Cooperators connect to others in order to create more; free riders connect in order to leach off those who create; and punishers connect in order to drive away free riders.

Theories of punishment have been around for a while, but they have always gotten stuck on how such a behavior could have evolved in the first place.[6] In a world of free riders, a punisher must constantly exert energy to correct each and every transgression (which sounds exhausting). Punishers could easily be overwhelmed. But in

a world of disconnected loners, punishers would not have to punish anyone. Using this reasoning, we created our own model and
showed that small groups of interconnected, interacting cooperators
and punishers could coevolve in a world of people who otherwise
keep to themselves, and this pushes the whole population toward
higher overall levels of cooperation and connection.[7]

Hauert and his colleagues later extended our model, showing that
under general conditions it would create a mixture of people who
cooperate and free-ride and people who connect and disconnect.[8]
Moreover, they demonstrated that the population was frequently in
transition, meaning that we might expect to find different propor-
tions of individuals of different types at any given moment. Unlike
the models that predict too much cooperation or too little, the
extended model showed that cooperation would happen frequently
in a world where it was possible to monitor and punish free riders and where there was variation in the tendency of people to join
groups. In short, the model predicted two things: some people will
cooperate, and others will not; and some people will be well connected to the social network, and others—the loners—will not.

Long Live *Homo dictyous*

This kind of variation in human behavior is very hard for traditional
economists to explain. The standard way of thinking about human
beings in economics is that every person makes a decision without
considering the interests of others (except insofar as the interests of
others impinge on one's own). From this perspective, the reason any
cooperation arises is that the choices of the individuals involved are
incentive compatible: I scratch your back because I think you are
going to scratch mine. If we happen to be in a situation where it is
possible for you to walk away, then I will refuse to help you. And you
and everyone you know would also refuse if you were in my shoes.

In other words, there is no inherent variation in how much people are willing to help others. And social connections are meaningless.

Indeed, *Homo economicus* inhabits a brutal, dog-eat-dog world in which concern for the well-being of others does not exist. The expression *Homo economicus,* a slightly tongue-in-cheek construction, was first used at least one hundred years ago to describe a vision of our species as one that relies on self-interest to obtain the maximum personal good at the lowest possible cost. But even earlier, in 1836, philosopher John Stuart Mill was already propounding a model of "economic man" who "inevitably does that by which he may obtain the greatest amount of necessaries, conveniences, and luxuries, with the smallest quantity of labour and physical self-denial with which they can be obtained."[9] Implicit in this vision is that people are lazy and greedy but also rational and self-interested and self-directed. Such a model leaves no room for altruism. Further, it leaves utterly unexamined how people come to want what they want to begin with.

We propose an alternative. *Homo dictyous* (from the Latin *homo* for "human" and the Greek *dicty* for "net"), or "network man," is a vision of human nature that addresses the origins of altruism and punishment, and also of desires and repulsions. This perspective allows our motivations to depart from pure self-interest. Because we are connected to others, and because we have evolved to care about others, we take the well-being of others into account when we make choices about what to do. Moreover, by stressing our embeddedness, this perspective allows us to formally include in our understanding of people's desires a critical source: the desires of those around them. And, as we have seen, this applies to everything from our health behaviors to our musical tastes to our voting practices. We want what others to whom we are connected want.

Indeed, social networks can involve an endless flow of desires, propagate idiosyncratic disturbances, and give rise to tastes. We have basic desires (such as an interest in sex) that do not depend so much on the specific desires of those around us. But we also have many arbitrary desires—in music or clothes, for example—that are strongly influenced by others. In other words, some of our tastes may be for things that are made more desirable when others desire them. Once such beliefs arise, for whatever reason, they can spread and amplify within the network. Whether these beliefs, desires, or tastes arise as random irregularities or whether they have a more consistent and fundamental origin (for example, political ideology and religiosity have partly genetic bases), they nevertheless get magnified in the network, and they flow through it.

Who Killed *Homo economicus?*

Beginning in the 1970s, nontraditional economists began to test some of the most basic assumptions of their discipline, asking questions about cooperation and the origins of tastes. And many were shocked by what they found. In 1982, a group of economists developed a simple but clever experiment called the "ultimatum game" in which two players bargain over $10 given to them by the experimenter.[10] The first player is instructed to make an "offer" to the second player for how to divide the $10. He could offer to give it all away, keep it all, split it fifty-fifty, or divide it any way that it is possible to divide $10. The second player then gets to decide whether to accept or reject the offer. If she accepts, then they divide the money as agreed and both get to keep it. If she rejects the offer, then *neither* gets anything. Game over.

Now, traditional economists who assume that all actors in their models are self-interested *Homo economicus* reasoned as follows. The second player would rather get something than nothing. Even a penny is better than nothing. And the first player knows this, so he knows that the second player will accept any offer greater than zero. Economists therefore predicted that the first player would offer the second player a penny, keeping $9.99 for himself, and the second player would accept the offer.

But that is not what happened at all. In the initial experiments with American college students, economists found that subjects frequently rejected low offers. Offers of $2 were rejected about half the time, and lower offers were rejected even more frequently. Moreover, the people in the first player's shoes seemed to understand that this would happen, since unfair offers were rare. The most common offer was an exact fifty-fifty split, and, on average, the first player earned a little bit more than the second player, but not much more because rejected offers caused both players to lose everything. Given that the first player seemed to know what kind of offers would be accepted and rejected without even discussing this with the second player, his behavior was consistent with that of someone who was purely self-interested and trying to make the most possible money for himself. However, the second player's behavior was completely inexplicable. Why would she reject an extra one or two dollars from an anonymous person whom she had never met and would never cross paths with again?

The ultimatum-game experiments led to the development of another set of experiments involving a new so-called dictator game to study the effect of the second player's power to accept or reject an offer. In this game, the first player is given $10 and allowed to divide it between herself and the second player any way she likes. But now the second player can do nothing. The first player's offer is automatically accepted no matter what. Because the second player has no power, economists expected that the first player would keep everything, and many did. But many more did not. The average first player gave away about $2 to the second player. The results of this extremely simple experiment were difficult to explain if we thought of behavior as being driven purely by self-interest. People were literally taking money out of their own pockets and giving it to anonymous strangers.

The experiment further showed that there was inherent variation from person to person that might have an impact on how people behave. In our own work, we have shown that people who give more in the dictator game are more likely to vote, donate to a campaign, run for office, attend a protest, and help Hurricane Katrina victims.[11] They also score higher on psychological tests that assess how "humanitarian" a person is. And just as Hauert's evolutionary models predict, we find variation in the willingness of individuals to bear personal costs to help others. Some people care only about themselves. But the majority of us take other peoples' well-being and interests into account.

Situations like these are not entirely fanciful. In fact, real-life ultimatum games occur all the time, sometimes quite colorfully. For example, in 2006, a contractor named Bob Kitts was tearing down the walls of an eighty-three-year-old house near Lake Erie, Ohio, when he found two green metal boxes carefully suspended by wires inside a wall. In the boxes were white envelopes that bore the name "P. Dunne News Agency" and contained $182,000 in Depression-era bills. Saying that he was not raised to act any other way, Kitts notified the owner of the house, Amanda Reece, a high-school classmate who had hired him for the remodeling project. Pictures they took show them happy and grinning, sitting on the floor next to a coffee table covered with carefully laid-out piles of bills.

The trouble began not long afterward when they discussed how to share the money. Reece offered Kitts 10 percent. He wanted 40 percent. Perhaps if Reece had been familiar with the percentages the ultimatum-game experiments had revealed, she would have made a better first offer, and they would have been able to avoid a whole lot of trouble. Because they were unable to agree on a fair split, the information leaked, and the local newspaper, the *Cleveland Plain Dealer*, published a story about the find in December 2007, whereupon Dunne's descendants got involved. All twenty-one of them. Naturally, they wanted the money. In the meantime, Reece spent

some of it on a trip to Hawaii and claimed that $60,000 was sto-
len from a shoe box in her closet. After that, not much was left to
be shared among Reece, Kitts, and Dunne's descendants. A lawyer
for the Dunnes put the situation starkly: "If these two individuals
had sat down and resolved their disputes and divided the money, the
heirs would have had no knowledge of it. Because they were not able
to sit down and divide it in a rational way, they both lost."[12]

In Search of *Homo economicus* Around the World

The economic experiments we have discussed were carried out almost
exclusively on university campuses in the United States. Research
such as this typically attracts hungry undergraduates in need of cash,
which has led some to remark that we now know more about American college sophomores than any other animal. But when researchers
in other countries replicate the ultimatum-game and dictator-game
experiments with local college students, they find largely the same
results. The main notable exception is among undergraduates studying economics: they have been taught that the self-interested choice
is the most rational one, and so they are significantly less likely to
cooperate.[13]

Anthropologist Joseph Henrich wondered if the results would
generalize to people in the nonindustrialized world, so he tried the
games with the Machiguenga, an indigenous group he was studying
in the Peruvian Amazon.[14] Oddly, these people behaved much more
"rationally" than their counterparts in the developed world. Those
acting as the first player in the ultimatum game made lower offers, and
those acting as the second player tended to accept them, a result much
closer to what traditional economists had previously predicted.

The "Machiguenga outlier" quickly became a curiosity to a variety of researchers, and several decided to join Henrich for a three-day
conference on the subject at UCLA in 1997. These researchers

designed a study in which they would scatter to all corners of the globe to administer the ultimatum and dictator games to people in a variety of small-scale societies. The list of countries and peoples studied included Bolivia (Tsimané), Chile (Mapuche), Ecuador (Achuar, Quichua), Kenya (Orma), Tanzania (Hadza, Sangu), Indonesia (Lamalera), Mongolia (Torguud, Khazax), Papua New Guinea (Au, Gnau), Paraguay (Aché), Peru (Machiguenga), and Zimbabwe (Shona). The group also conducted experiments in the exotic locales of Ann Arbor, Michigan, and Brentwood, California, as controls. Subjects were offered high stakes in most cases: they bargained over a day's wages or more to ensure that the decision would be taken seriously.

It turned out that the Machiguenga were not alone. From one society to another, there was a wide range of variation both in the initial offer and the tendency to reject unfair offers. The Ann Arborites of Michigan performed as previously observed among university students, with average offers by the first player of about 44 percent in the ultimatum game. The Hadza of Tanzania and the Quichua of Ecuador offered much less than university students, about 27 percent on average. Meanwhile, the Lamalera of Indonesia and the Aché of Paraguay offered somewhat more (58 percent and 51 percent, respectively).

The anthropologists involved in the study were curious why some groups would offer more than others, so they studied several aspects of each society, including the nature and origin of the local language and the peoples' relationship to the physical environment. Some groups lived in forests, others on the plains, and others in deserts. Some groups were hunter-gatherers, others were shepherds; still others were small-scale farmers. Some groups lived sedentary lives, staying in one place, while others were nomadic, roaming widely. Their societies varied in complexity too, with some based on families, and others based on groups of families, tribes, or villages.

But the variables that seemed to be most closely related to differ-

CONNECTED

ences in group behavior were explicitly social. One of these variables was *anonymity,* a measure of how frequently each group interacted with strangers. For example, the Achuar of Ecuador hardly ever saw strangers, but the Shona of Zimbabwe encountered them all the time. The groups also varied in how frequently they engaged in market transactions. The Hadza foragers in Tanzania were almost completely self-reliant, and they therefore had little interaction with markets; whereas groups like the Orma of Kenya frequently bought and sold livestock and also worked for wages from time to time.

The researchers found that the groups who were in greater contact with strangers were also more likely to engage in what they call *prosocial behavior.* That means they cooperated with others in the ultimatum game by making higher offers, but they were also willing to reject low offers. In other words, as groups expand their networks beyond the family unit, they appear to behave less and less like *Homo economicus* and more like *Homo dictyous.* The economists' simplification of a person who would offer little or nothing to his counterparts would thus seem to apply only, if ever, to isolated individuals bereft of social interaction, a state that cannot easily be found even in the most remote parts of the globe and that has never really characterized the human condition.

Learning from Twins

Over the course of human history, we have gradually come together to live in clans and villages and then towns and cities. We have become progressively more connected, even to strangers. And, as we will see in chapter 8, this development has moved from the real world into cyberspace. But we are not all interchangeable cogs in a machine. Human beings clearly exhibit strong differences in both their tendencies to care about others and their abilities to connect. While our life experiences may influence whether we help strangers or make friends, it is also clear that we carry these differences deep within us from a time long ago when our ancestors were first learning to live together in small groups.

Anthropologists think we began hunting large game about half a million years ago. This was quite an innovation because big animals were plentiful at the time, and a single animal could sustain a group for several days. To be successful, however, our human ancestors had to cooperate. And we have already seen that participating in group activities yields a diversity of coexisting strategies. Some people go it alone, while others contribute to the group. Some people free-ride on the efforts of others, while others willingly punish free riders. If this has been the human condition for hundreds of thousands of years, then maybe our network behavior is not simply the product of markets or of a growing population density. It could be that we have evolved genetically to adapt to the risks and opportunities of cooperating in groups. If so, then we should expect to find evidence that cooperative behavior and our tendency to connect are written in our DNA.

To test this theory, we started in an unlikely place. In the summer of 2006, we traveled with our colleague Chris Dawes to a sleepy town called Twinsburg in the middle of rural Ohio. An annual festival is held there, complete with carnival games, hay rides, and funnel cake; in many ways it seems like any other country fair. But this festival is different. It started in 1976 when some of the citizenry thought it would be nice to honor the town's name by setting aside one day during their bicentennial festivities to celebrate twins. Thirty-seven pairs showed up that first year, and the idea quickly caught on. By 1985, more than one thousand pairs of twins were visiting the Twins Days festival, and they continue attending at this rate today, making it the largest annual gathering of twins in the world.

Each year, things start off with a wiener roast for the twins and their families, followed by events that are open to the public like the "Double-Take Parade," where the twins march by twos from the center of town to the fairgrounds, and a series of twin contests, including several "most alike" and "least alike" competitions. And tucked away to one side is another spectacle, albeit more subdued. Dozens of researchers from around the world travel to Twinsburg each year to study the health and behavior of twins. In tents, researchers ask twins questions about their childhood, take saliva and blood samples, administer vision and hearing tests, and even do dental exams. Each of the studies pays a little money that the twins can use during their time at the fair. In fact, there is usually a rush right before lunch when hungry volunteers descend on the tents to earn funds for their funnel cake.

Researchers like to attend the festival because twins provide a unique opportunity to study genes. Some pairs of twins are identical: they share exactly the same variants of every single gene in their DNA. Other pairs are fraternal, sharing only half of their genes on average. Differences in genetic similarity turn out to be a powerful natural experiment, allowing us to estimate how much genes influence a given trait. For example, identical twins almost always have the same eye color, but fraternal twins often do not. This suggests that genes play a role in eye color, and in fact geneticists have identified several specific genes that are involved. In the same way, scientists can estimate the role genes play in any other trait by comparing the similarity of identical twins to the similarity of fraternal twins. If there is no difference, then genes play no role. If there is a difference, then the magnitude of the difference gives a clue as to how much genes are involved.

The twin-study method is not without its critics. Some people argue that if twins self-identify as being identical, then they will strive to become more similar, their parents will treat them the same, their friends will treat them the same, and so on. Hence, they would resemble each other for social rather than genetic reasons. If identical twins dress alike and eat the same foods and enjoy the same movies, it might be because their social environment encourages them to be similar. But this claim has been tested in ingenious ways. Occasionally, a pair of twins will mistakenly think they are identical and so will all their friends and family members, but a simple genetic screen shows they are not. If the social environment is really what makes identical twins more similar, then these misidentified identical twins should be just as similar as real identical twins. Yet, when scientists tested a variety of characteristics (intelligence, personality, attitudes, and so on), they found that the misidentified twins are only as similar as fraternal twins. That means that it is the genetic state of being identical and not the self-perception of being identical that drives similarity.[15]

At Twinsburg, we used the twin-study method to measure how much genes play a role in a simple test of cooperation called the "trust game." In this experiment, we paired each twin with a person she did not know and assigned them the roles of player 1 and player 2. We gave player 1 $10 and asked her to choose how much money to give to player 2. We also told both players that each dollar sent to player 2 would be tripled. For example, if player 1 gave away all $10, then player 2 would receive $30. Player 2 was then asked to choose how much money to send back to player 1 (but this time it was not tripled). So if player 2 received $30 and wanted to split it fifty-fifty, then he could send back $15 to player 1 and keep $15 for himself. Player 1 would earn an extra $5 as a result.

This game is called the trust game because the first player's decision indicates how much she trusts the second player to return some of the money she is giving away. The more she gives, the more she trusts the other person. By the same token, the second player's decision indicates how "trustworthy" he is. The more he returns, the more he is reciprocating the first player's initial generosity. Higher values of both trust and trustworthiness indicate more cooperative, prosocial behavior.

Over the course of two summers at Twinsburg, we had about eight hundred twins play this game, each of them with a person they did not know. We then compared the identical twins to the fraternal twins. The results indicated that genes significantly influenced both trust and trustworthy behavior. And in one of those happy accidents that occur so often in science, by chance we traded e-mails with David Cesarini, an economist from MIT who was doing exactly the same thing with several hundred sets of twins in Sweden. Amazingly, he had observed nearly identical results in his own study. So we joined forces, and both studies were published side by side.[16]

Since then, David Cesarini and Chris Dawes have found that genes influence behavior in the dictator game and the ultimatum game that we described earlier. What this means is that cooperation, altruism, punishment, and free-riding are written into our DNA. There can be no doubt that our life experiences have a big impact on all of these characteristics, but for the first time we had found evidence that diversity in these social preferences may be at least partly a result of our genetic evolution.

Networks Are in Our Genes Too

Human sociability and social networks have ancient genetic roots: apes form social ties, hunt in groups, maintain enduring social bonds, and derive advantages in terms of how long they live and how well they reproduce from these ties. But humans take these traits to a whole new level. The tendency to form social unions beyond reproductive ones is biologically encoded in humans: we seek out friends, not just mates. And through our research, we have discovered that genes also play a role in more complicated aspects of social-network structure. In fact, genes have a big impact not just on whether we are friendly but also on where exactly we land in the vast social network that surrounds us.

To explore the role of genes in social networks, we analyzed 1,110 twins drawn from a national sample of 90,115 adolescents in 142 schools (this is the same Add Health data set used to study adolescent sexual behavior discussed in chapter 3).[17] These students had been queried about their friendship networks, and the entire friendship network at each school was known, as well as the student's precise position in it. We started by studying the most fundamental building block of a human social network: the number of times a person is named as a friend. We found that genetic factors were very important, accounting for about 46 percent of the variation in how popular the kids were. On average, a person with, say, five friends has a different genetic makeup than a person with one friend.

This result by itself was not too surprising. We already knew, for example, that facial symmetry is heritable and is associated with beauty, and that this could help to explain why some people naturally attract more friends. But what was surprising was that even higher-order structural aspects of a person's network position appeared to be under genetic influence. Your genes affect not just how many friends you have but also whether you are located in the center or at the periphery of the network. On average, people located in central parts of the network have a different genetic makeup than those located at the periphery.

We also studied the effect of genes on how interconnected your friends are. Recall that *transitivity* denotes the probability that any two of your friends are also friends with each other. People with high transitivity live in densely clustered cliques where everyone knows everyone else. People with low transitivity, in contrast, tend to have friends in several different groups. Such people often act as bridges between completely different groups of people. In our study, we found that transitivity is significantly heritable, with 47 percent of the variation explained by differences in genes. Therefore, on average, a person with five friends who know one another has a different genetic makeup than a person with five friends who do not know one another.

One bizarre implication of this finding is that if we knew which genes were involved in transitivity, we could use that information from one person to predict whether two other people were likely to become friends. If Tom, Dick, and Harry are in a group, this means that Tom's genes affect whether Dick and Harry are friends!

How can this be? Recall that in chapter 3 we discussed how people commonly meet their partners through others. Introductions are a key feature of human life. But not everyone goes around getting their friends together. In fact, some people go to great lengths to prevent connections. In a memorable episode of the television show *Seinfeld*, the character George Costanza worried so much about his two sets of friends meeting each other that he claimed it would "kill independent George." That episode hit the zeitgeist, spawning a "Worlds Collide Theory," which, according to the Urban Dictionary, "states that a man must keep his personal side (i.e., friends) separate from his relationship side (i.e., girlfriend). Should the two worlds come into contact with each other (by means of his girlfriend becoming friends with his friends), both worlds blow up."[18]

Based on our findings about the heritability of social-network attributes, we developed a mathematical model for how social networks form. This "attract and introduce" model is built on two simple assumptions. First, some people are inherently more attractive than others, whether physically or otherwise, so they are nominated more often as friends. Second, some people are inherently more inclined to introduce new friends to existing friends or to attempt marriage matches among their friends (and so these people indirectly enhance their transitivity).

As a result of these two behaviors, we also affect how central we are to the network. Diversity in these choices (whether conscious or not) yields an astonishing variety of locations we can inhabit within our social network, and this in turn has important implications for our lives. If genes influence whether we are in the middle or at the periphery of a social network, they can also affect how rapidly we get a piece of gossip (the center is better) or how likely we are to become infected with an epidemic disease (the center is worse).

The variation in whether certain positions are indeed better or worse, depending on circumstances (acquiring information, avoiding a germ), helps explain why not every position in human social networks is the same. If it were always beneficial to have a certain number of friends and a certain number of connections between our friends, then our social world would look like a boring, predictable lattice or the atoms locked in a salt crystal. We would all have the same kind of network.

Traits that are always adaptive tend to reach what geneticists call *fixation* in the population: in the long run, everyone becomes the same. But when there are conflicting pressures — under some circumstances, a trait is beneficial, but under others, it is not — then it is possible to maintain diversity in the population in the face of natural selection. If it were to our advantage to all be the same height, then we would be, because long ago evolution would have culled the very tall and the very short. Likewise, if only one kind of social network and location within it were optimal, we would all have the same kind of social networks and be located in identical positions within them.

There are probably many reasons for genetic variation in the ability to attract or in the desire to introduce friends. More friends may mean greater social support in some settings or greater conflict in others. Having denser social connections (greater transitivity) may improve group solidarity, but it might also insulate a group from beneficial influence or information from people outside the group. But most important, social networks may serve the adaptive function of transmitting emotional states, material resources, and information between individuals.

Some traits that we have shown to spread in social networks also appear to be heritable (such as obesity, smoking behavior, happiness, and political behavior). That means that a full understanding of these traits will likely require a better understanding of the factors that mediate the influence of genes on social-network structure as well as how the patterns that we see in our networks today might have evolved.

Lonely Is the Hunter-Gatherer

Genes may influence our tendency to form social ties through their regulation of our emotional states. When core needs for intimacy, love, and social connection are not met, people often experience a feeling of loneliness. Feeling lonely is not the same thing as being alone, however, and there is often a disjunction between a person's psychological sense of disconnection and their more objective, sociological placement in a social network. Feelings of loneliness can arise from the discrepancy between our desire for social connection and our actual social connections.

Because there is a survival benefit for those who find social relationships helpful, this would promote genetic selection of individuals with such emotional responses. Yet, we have already seen in the evolutionary models we described earlier that it is possible for loners (people who opt out of group activities) to survive. Psychologist John Cacioppo and his colleagues make this mathematical prediction more concrete, arguing that, many thousands of years ago, hunter-gatherers struggling to survive in times of undernourishment may have considered *not* sharing their food with their family—that is, they may have embraced the loner strategy.[19] Individuals who did not feel any loneliness in the absence of family or friends may have been more likely to survive, but their offspring would have been less likely to survive the period of undernourishment given the lack of food.

Conversely, individuals inclined to share food with others may have reduced their own chances of survival but increased those of their offspring, suggesting that no single strategy is always best. As a consequence, a diversity of feelings about being connected and sharing with others were able to evolve, leading to heritable differences in loneliness in adults. And this is exactly what behavioral genetics has revealed. Recent studies of 8,387 adult twins from the Netherlands

Twin Register confirm that about half the variation in whether a person feels lonely or not depends on his genes.[20]

We do not know whether the same genes that affect loneliness also affect a person's social networks, but the results are suggestive. In chapter 2 we saw that emotions can spread from person to person to person and that people who feel lonely are more likely than others to disconnect from the network. As a consequence, it is possible that genes regulate the structure of our social networks via their effect on our moods. If it pays to be a loner when everyone else in the population is cooperating, free-riding, or punishing, then natural selection might favor those genes that promote the feeling and spread of loneliness. But the advantages of being connected mean that there is an upper limit on the number of people who could benefit from going it alone.

Voles, Macaques, Cows, and Senators

Social connection is a complex phenomenon that likely involves hundreds of genes and is probably influenced by countless gene-environment interactions. But some specific genes appear to have a truly remarkable effect. Scientists have recently shown that a single gene variant distinguishes mating and parenting behavior in small mouselike mammals called voles.[21] Male prairie voles are paragons of monogamy, attaching to their first mate for life and taking care of their kids. Male meadow voles, in contrast, are much more promiscuous and less likely to care for their young. This stark difference in mating behavior suggests that evolution does not always yield behavior that we might consider moral in humans—sometimes it favors lust and the deadbeat dad. But more important, it shows what a big difference even a single gene can potentially make in the way animals connect to others. And this has led to similar research in humans, which has shown that people with a certain variant of a related gene give significantly more of their money to anonymous recipients in the dictator game—that is, they are more prosocial.[22]

Given the role of genes in social networks and cooperative behavior, and the fact that other animals seem to be similarly affected, it might seem that human social networks are nothing special. Humans do share much in common with other social species. For example, primate social networks (in chimpanzees, gorillas, orangutans, and so on) are built on grooming—the activity most of us have seen in nature documentaries where apes appear to be foraging through the hair of their confederates. This close-proximity activity allows one individual to get to know the other, his behavior, health, tendency to violence, willingness to reciprocate, and so on. Primates also use grooming to form alliances, and animals' willingness to come to mutual aid has been observed to be directly proportional to the amount of time they spend grooming each other.

Primatologist Jessica Flack and her colleagues recently used a "knockout" study to show that removing important individuals from a group of pigtailed macaques significantly changed the structure of interactions in the network of grooming and play.[23] This change in network structure, in turn, yielded decreased cooperation and greater instability in the group's behavior. For an analogous example in humans, consider what happens when you take a teacher out of a middle-school classroom or when a referee ejects one of the eleven players on a soccer team. Both the person and all their ties are lost, and things do not work as smoothly.

The link between cooperation and networks does not appear to be uniquely human. Sociologists Katherine Faust and John Skvoretz tackled the issue of human uniqueness head-on in their own study of forty-two social networks sampled from fifteen different species, including chimpanzees, three kinds of macaques, patas monkeys, vervet monkeys, cows, hyenas, highland ponies, red deer, silvereyes, sparrows, tits, and human beings. Among humans, they looked at networks of managers, monks, and even U.S. senators. Within this enormous variety of network structures, there were some important similarities. They found that the type of relationship in the network was much more important than the identity of the species for making good predictions. For example, grooming relationships appeared to be very similar across species. In fact, the model that best predicted the network structure of U.S. senators was that of social licking among cows.[24]

However, primates have cognitive abilities that are especially well suited to making sense of social information. They can recognize individuals; distinguish kin from nonkin; assess and compare the value of resources and services offered by others; remember past interactions with particular group members; discriminate between cooperators and free riders; and assess the relative desirability of prospective rivals, mates, and allies. They also have "third-party knowledge" about the relationships between other members of their groups. For example, in one set of experiments with vervet monkeys, the investigators played tape recordings of the screams of baby monkeys, and the adults in the group looked not toward the screams but preferentially toward the female they knew to be the mother of the baby whose voice was being heard, the same way human adults hearing a baby crying on an airplane would. Crucially, primates are also able to manipulate their social network in response to events such as loss of a partner. For example, females in one species whose close kin member died responded by spending more time grooming other social contacts and expanded the number of individuals with whom they interacted. Humans behave similarly: a recently bereaved woman may join a variety of new groups in order to make new friends or to find a new spouse, and her friends may band together to help speed her recovery from grief.

Given the similar (if simpler) behavior in primates, it should not be surprising to find that many human social-network behaviors are hardwired. Humans, of course, manipulate and interact with social networks in ways more complex than other animals, and in the process our social networks can change. Paradoxically, as we will see in chapter 9, while networks depend on their members, they also are inherently stable, and new members can come and go as part of the normal process by which networks evolve and survive.

A Brain for Social Networks

So far, we have drawn on evidence from remote human societies, our primate cousins, other social animals, and even our genes to support the idea that social networks are an ancient part of our genetic heritage. But unlike other animals, human beings cooperate with unrelated individuals in enormous and complex societies. And to navigate this complexity requires some special abilities that only humans have. In particular, our brain seems to have been built for social networks.

Compared to other species, humans have unusually large brains and unique cognitive capacities, ranging from language to abstract mathematics. Evolutionary biologists and physical anthropologists have developed a variety of explanations for the origin and function of the human brain. The *general-intelligence hypothesis* posits that larger brains have allowed humans to perform all kinds of cognitive operations (ranging from superior memory to faster learning) better than other species. The *adapted-intelligence hypothesis* posits that particular mental faculties evolved in response to particular environmental challenges. For example, caching birds that store food in a variety of locations have terrific memories, and social insects have complex communication skills.

An alternative theory about our big brains that has recently gained momentum is the *social-intelligence hypothesis,* which stresses the special challenges posed by living in close proximity with others and confronting a complex social environment that demands constant cooperation or competition. In other words, this is essentially a theory about networks. It suggests that humans are "ultrasocial," with skills ranging from language to abstract reasoning to empathy and insight that are adapted to a highly social environment. It also suggests that skills have evolved in humans to create and form social groups, to manipulate the social world, and to shape the architecture of the social networks in which we are embedded. These groups can be discerned both at a large scale (where they constitute whole cultures, with their own languages and artifacts) and at a small scale (involving, say, just kin-group interactions). Once humans form particular social groups with particular social-network ties, they can then transmit their knowledge to others near and far.

At some point in evolutionary history, primates applied the skills initially developed for finding a mate and maintaining a stable reproductive union to relationships that were not reproductive. As primatologists Robin Dunbar and Susanne Shultz argue, the everyday relationships of many primates entail a type of attachment that is found only among pairs of individuals bonded for reproduction in other species. Primates of the same or opposite sex could form long-term, stable relationships to their mutual benefit. But each of these relationships brings with it the potential for many more relationships, since every friend is also potentially connected to every other friend. There is only one relationship between two people, but there are three possible relationships between three people, six between four people, ten between five people, and so on. Since the number of possible relationships grows exponentially with group size, it probably takes a big shift in cognitive capacity to keep up with all the drama of a full social life.

More direct evidence of the social-brain hypothesis comes from the use of functional MRI to study the neural correlates of social decision making. Neuroscientists have found that we use a very large part of the brain called the *default-state network* to monitor social interactions, and we even seem to have expanded the use of these brain regions to think about coalitions and conflicts in modern-day politics.[25] Biologists have also found that color vision, which occupies about two-thirds of a human brain's capacity, is optimally tuned to detect differences in skin color. This might be so that individuals can discern emotional states in other members of the same species. And, intriguingly, the species that have this ability also have little facial hair (like humans, sometimes called the "naked ape"), suggesting that color vision coevolved with the need to be able to see the faces of other group members and assess their moods.[26] Thus, over millions of years, our social life has affected not merely our ability to monitor others and make decisions; it also may have changed the very way we see the world.

Anthropologist Michael Tomasello and his colleagues have taken the social-intelligence hypothesis a step further and argued for a variant called the *cultural-intelligence hypothesis*, positing that higher cognitive functions are based on a whole complex of social skills. He writes: "There should be an age…before children have been seriously influenced by written language, symbolic mathematics, and formal education at which humans' skills of physical cognition (concerning things such as space, quantities, and causality) are very similar to those of our nearest primate relatives but at which their skills of social-cultural cognition (specifically those most directly involved in cultural creation and learning, such as social learning, communication, and theory of mind) are distinctively human."[27]

The cultural-intelligence hypothesis finds support in experiments on chimpanzees, orangutans, and two-and-a-half-year-old human children using an omnibus Primate Cognition Test Battery—an IQ test of sorts—that offered treats for a variety of tasks, ranging from locating a reward, tracking a reward after invisible displacement, discriminating quantity, understanding that things can change in appearance, understanding functional and nonfunctional tool properties, and so on. The IQ tests administered in this head-to-head comparison of human toddlers and adult primates included two in which the toddlers had a particularly strong advantage: following the direction of an actor's gaze to a target, and reading the intentions of others.

These tasks showed that even at this young age, when physical cognition in children is similar to that of apes, humans outperform them on tasks in the specifically social domain.

In short, the human brain seems to be built for social networks. Over time, evolutionary selection has favored larger brains and greater cognitive capacity to respond to the demands of a more complicated social environment. Individuals living in social networks confront a set of cognitive challenges not faced by solitary individuals or by those in disconnected groups. These challenges arise from the need to understand others and to cooperate with them, as well as to occasionally act altruistically for the benefit of the group. A bigger brain is needed to avoid self-destructive aggression, to hunt down a mastodon, and to avoid getting voted off the island.

Connected to a Higher Power

There is increasing evidence that religion and the inclination to form social networks are both part of our biological heritage and that the two may be related. Religion is one means of integrating people into a collective. A belief in God can have relevance to social networks in a direct way: God can actually be seen as a part of the social network. This involves not just the personification of a deity but the addition of a deity into the social fabric.

One way to make social networks stable is to arrange them so that everyone is connected to a node that can never be removed. There would then be a short path from each person to every other person through this particular node. But even the most popular person in a society could not fill this role since, realistically, a single individual cannot be connected to absolutely everyone. And even if someone could be so connected, the effect on the network would not be permanent because humans are mortal.

But such considerations do not apply to God. If God were seen as a node on a network, large groups of people could be bound together not just by a common idea but also by a specific social relationship to every other believer. People could perceive a specific social tie to others, and everyone would be one degree removed from everyone else. People who felt a connection to God would have a way of feeling connected to others, because through God everyone is a "friend of a friend."

This is not merely an abstract idea. People often see their networks in just this way. For example, in the early 1980s, psychologist Catalin Mamali became interested in how people perceive their relationships to others and how they form mental maps of their relationships. She developed a method to ascertain these mental maps by asking people to identify others with whom they interact, and to draw their relationships in a kind of network graph.[28] Her subjects were told to think of people they were "close to" and who were "highly significant" to them and to draw their interconnections. Specific examples of such people were provided: parents, siblings, spouses or intimate partners, children, best friends, friends, and neighbors. Very unexpectedly, however, quite a few people following these instructions chose to include God as a node on their networks and even explicitly made God connected to everyone. The figure below is one such "social autograph" from an eighteen-year-old college student.

The idea that God can be personified and seen as part of a human social network is further supported by the fact that people tend to have an increase in religious belief after the death of a loved one: it is as if a connection to God is strengthened when one loses a connection to people. People might also turn to God in order to support a belief in an afterlife; this belief can sustain the hope of being reconnected with others who have departed. And the very fact that most gods are personified is consistent with the idea of God's inclusion in social networks and the sense that many people have, and that religions foster, that "God is among us."

There is a tendency for socially isolated people—and not just

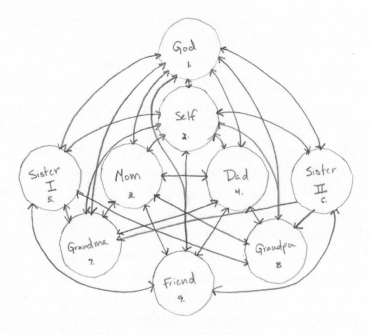

"Social autograph" of relationships identified by an eighteen-year-old college student (subject 1987SA) showing how she feels she is connected to important people in her life. (Courtesy of C. Mamali.)

bereaved individuals—to anthropomorphize the world around them, from mermaids in the ocean to faces on rocks. Studies by psychologist John Cacioppo and his colleagues show that people who are experimentally induced to feel lonely will alter their mental state in order to personify, and feel connected to, "gadgets, greyhounds, and gods." Religion is the opiate of the people, Karl Marx famously noted, but it turns out that it is the opiate of disconnected people in particular. Cacioppo and his colleagues administered personality tests to ninety-nine people (fifty of whom believed in God and forty-nine of whom did not) and then randomly assigned them to receive one of two possible interpretations of their performance on the test, regardless of how they actually performed. One interpretation was: "You're the type who has rewarding relationships throughout life" (that is, you will be connected); and the other was: "You're the type who will end up alone later in life" (that is, you will be disconnected).

The subjects then rated the extent to which they believed in ghosts, angels, spirits, God, and so on. Not surprisingly, subjects who reported believing in God before the start of the study reported a strong belief in these supernatural agents. However, regardless of their belief in God, those who were told they would end up disconnected reported an increased belief in supernatural agents.

Lest we think these findings are specific to God, these investigators also did an experiment in which subjects were encouraged to believe that they were more connected or more disconnected, and then they were asked their feelings about pets. The investigators reasoned that if disconnection made people seek out and personify agents in their environment in order to regain a sense of connection, this sensibility should also apply to pets. In fact, this is what they found.[29]

Making people feel disconnected did not, of course, turn atheists into deeply religious people, but it did nudge people in the direction of believing in God. To the extent that the saying "there are no atheists in foxholes" is true, we can add a wrinkle: this is especially likely to be true if you are all alone in the foxhole.

Our claim is not that God is really a part of social networks, of course, but, rather, that one way to understand religion is to study its role in the function of social networks. Religious sensibilities are partially hardwired in our brains, and they are related to our desire for social connection to others, not only a spiritual connection to God. A key function of religion, in other words, is to stabilize social connections.

Investigations of the fundamental ways in which the mind works further support this idea. For example, functional MRI studies show that during religious feelings and altered states of consciousness, the parts of the brain that regulate the sense of self in time and space actually stop functioning. This contributes to the sensation that "all are one" and may help us overcome a built-in rigidity in the way we perceive our position with respect to others.[30] In essence, the brain is fooled into believing that social boundaries do not exist or, equivalently, that everyone is connected to everyone else. Thus, people who would be willing to do something to help their friends suddenly have many more friends and may be more motivated to act on their behalf. In this way, a religious movement can bring together disparate groups of people to achieve a common goal, from caring for the poor to building great structures to, alas, launching wars on rival groups.

How Many Friends Can You Have?

As we will see in chapter 8, people sometimes claim to have hundreds of friends in their online social networks. While the human brain is designed to cope with large social networks, our capacity for friends is not, in fact, unlimited. As a key exponent of the social-brain hypothesis, Robin Dunbar has argued that the evolution of larger social groups among primates required and benefited from the evolution of a larger neocortex (the outer, thinking part of our brain), and that managing social complexity required and benefited from the evolution of language. In a celebrated 1993 paper, which was published along with the comments of more than thirty other scientists, Dunbar examined the relationship between brain size and group size in a variety of primates and by extrapolation posited that the expected size of social groups in humans, based on our big brains, should be about 150. This has come to be known as *Dunbar's number*.[31] Animals cannot maintain the cohesion and integrity of social groups larger than a size governed by the information-processing capacity of their brains. And humans have both the largest groups and the largest brains.

But what do we mean by *groups*? We clearly do not see clusters of humans grooming each other on the street. In primates, it is easy to define the boundaries of a group—who is in and who is out. But in humans, there are groups as large as tribes and nation-states and as small as families and villages. Dunbar argues that a group is the maximum number of humans in which every member knows every other member, knows whether they are friendly or hostile, and knows the relationships among them. This is the number of people whom you recognize and with whom you can maintain a stable, coherent relationship, which Dunbar defines as "one that can be picked up again on meeting after an absence without any need to reestablish where you stand." According to Dunbar, an informal definition might be "the number of people you would not feel uncomfortable about joining, uninvited, at a chance meeting in a bar."

But how do we decide what the natural condition for humans is in order to test the prediction? Here, Dunbar creatively employed data from a variety of sources. For example, a survey of the ethnographic literature looking at all documented censuses of hunter-gatherers showed three types of social groupings: the "overnight camp," the "band or village," and the tribe. The mean sizes for these were 38, 148, and 1,155 people, respectively. Hence, remarkably, the size of a band or village matched Dunbar's number. He also noted that the size of Neolithic villages in Mesopotamia in the period 6,500 BCE to 5,500 BCE fell in the range of 150 to 200 inhabitants. A more subtle point from these studies is that while tribe and overnight-camp size varied widely, the size of a band or village was much more consistent, suggesting that it is a more fundamental grouping.

Dunbar also studied the case of the Schmiedeleut Hutterites, a fundamentalist group who live and farm communally in South Dakota, North Dakota, Minnesota, and in Manitoba, Canada. The Hutterites explicitly regard a community of 150 to be the limiting size, and they make arrangements to split into two groups as they approach that number. They note that this is the maximum group size that can be maintained through the workings of peer pressure alone and that for larger groups a police force and hierarchical structure would be required.

For centuries, the size of fighting units in armies has been about 150 men. The basic unit in the Roman army (the *maniple*) was 120 men, and the mean size of the company, the analogous unit in modern armies, is about 180. These numbers suggest an upper limit on the size of a group in which members can work together as a coordinated team and know the strengths, weaknesses, and reliability of their brethren. One can even imagine that warfare presents a particular kind of evolutionary selection pressure, and that armies across the centuries have arrived at this working size by empirical observations of what size group is most likely to survive.

Interestingly, group size in modern armies hasn't changed, even though modern telecommunications would seem to facilitate larger group coordination. This suggests that communication is not the crucial factor. More important is the human mind's ability to track social relationships, to form mental rosters that identify who is who, and to form mental network maps that track who is connected to whom and how strong or weak, cooperative or aggressive, those relationships are.

Groom Your Friends, or Just Talk to Them?

Dunbar's assessments of the relationship between brain size and group size required some further novel claims. He predicted how much time primates would spend grooming each other in order to maintain the cohesion of the larger groups made possible by bigger brains. Dunbar estimated that, for the 150-person group size predicted for modern humans, we would have to spend 42 percent of our time grooming each other! He argued that language emerged among humans, in part, to replace grooming. Basically, language is a less yucky and more efficient way to get to know our peers since we can talk to several friends at once but only groom them one at a time. In fact, in a conversation with a small group, we can assess the behavior, health, aggressiveness, and altruism of several individuals simultaneously. Plus, we can talk to someone else while engaged in another activity, like foraging for food in the refrigerator.

This is actually quite a radical idea. Until relatively recently, it had been thought that language had evolved in order to facilitate an exchange of information—about the location of predators or prey, for instance—or as a secondary consequence related to our development of tools. But the social perspective on language suggests that it evolved as a means of maintaining group cohesion. Just like emotions, language is key to the acquisition and manipulation of social information about other members of our species. A strong indicator of this fact is that most conversation is light on intellectual content, and it certainly is not focused on complex ideas about the environment we inhabit or even about culture or economics. After all, how often—unless we are ten-year-old boys—do we talk about predators or navigation, let alone particle physics or Homer?

Dunbar estimates that language would have to be 2.8 times more efficient than grooming in order to sustain the group size seen in humans. Hence, he estimates that human conversational groups should be about four people: one speaker and 2.8 listeners. But how can we quickly get a sense of the size of groups people form for conversation? One strategy was to assemble data regarding restaurant reservations. Over a ninety-eight-day period at Novak's Restaurant in Brookline, Massachusetts, in 1968, for example, 3,070 groups made reservations. While there were peaks at two and four people, the average party size was 3.8. Everyday experience with dinner parties supports this result, of course. At large dinner parties, conversational cliques tend to form among four people. Another adventurous scholar—apparently without ulterior motive—observed the size of swimsuit-clad groups on a public beach and found a similar number.

Finally, Dunbar notes that the emergence of language had a further, unintended consequence. After language evolved to handle social interactions on a small scale, we were free to use it to do things like write poetry—just as feathers evolved to keep prehistoric reptiles warm but they wound up using the innovation to fly. In addition, and more crucially, we could use language to handle social interactions on an enormous scale, allowing us to interact in groups such as tribes and nations.

Language permits this transition to large-scale interactions in at least two ways. First, it makes it easy for us to categorize people and interact with them as types rather than as individuals; for example, instead of knowing every policeman (and grooming them), we interact with them in a stylized way ("What seems to be the problem, Officer?"). Second, it allows us to teach others how they should behave with certain types of individuals ("Stand at attention when the teacher comes into the class, students."). Hence, a person interacting with another person for the first time knows how to do so, without having the benefit of prior knowledge of that particular individual.

The tendency to form ties and to live out our lives in social networks has had an important effect on our development as a species. Social networks encouraged rapid growth in the size of our brains, which allowed us to acquire language and to become the dominant species on the planet. At the same time, these biological changes have given us the capacity to cooperate in large groups, even with complete strangers, to create magnificent and complex large-scale societies. Our connections—partly shaped by our genes but also profoundly influenced by our culture and our environment—are made and remade every day. We choose our friends, develop cultural norms about social order, make and obey rules about whom we can date or marry, enforce ideas about reciprocity, and react to events affecting people around us, in part because we are equipped with empathy. And as we will see in the next chapter, our ability to manipulate and participate in networks is highly relevant to the new opportunities and challenges that we are facing in our hyperconnected world. As we take our real-world networks online, we carry with us the tools wrought by natural selection to make a new world that nature has never before seen.

CHAPTER 8

Hyperconnected

very month, eleven million people around the globe play a game on the Internet known as World of Warcraft. It is a "massively multiplayer game" involving so many players that if it were its own country, it would be larger than Greece, Belgium, Sweden, and nearly 150 other nations. In this game, people adopt an online persona, known as an avatar, who inhabits a virtual world and interacts with other players in the game. This avatar has a vivid three-dimensional appearance that is customizable, and it acquires possessions, powers, and even pets over the course of play, which can last many months. Within this game, people form friendships, have sustained interactions as groups, communicate using instant messaging, collaborate to achieve shared goals, engage in economic transactions, and fight one another in complex battles. The avatars live in different realms of the virtual world, and when they "die" during combat or other activities, they are automatically returned to their homes, whereupon they happily come back to life and resume play.

Sometimes, however, things run amok. On September 13, 2005,

the game developers opened up a new area for advanced players, one inhabited by a massive, powerful winged serpent called Hakkar. Hakkar was equipped with a number of weapons and capabilities, among which was a contagious disease called "corrupted blood" that he could spread to his enemies. When one of his adversaries was infected, other nearby opponents also became infected. To the strong players who had banded together to fight Hakkar, this infection was intended to be a minor hindrance that made combat more challenging. Once Hakkar was dead, players could leave the area and the contagion would stop.

The programmers at World of Warcraft thought this was a pretty neat trick to challenge their players. But the players responded to the contagion in an unanticipated way. Rather than continuing the fight against Hakkar until they died of corrupted blood, some players used a teleport capability to transport themselves to another area of the game. As a result, the infection spread widely throughout the entire virtual world, not just among the players confronting Hakkar. What was intended to be a minor inconvenience to powerful players in a localized area—something like a cold in a healthy adult living in a small town—instead inadvertently became a worldwide epidemic in the game, rapidly killing hundreds of thousands of weaker players.

As players returned to their virtual homes, they spread the infection far and wide, including to the densely populated capital cities. In addition, through another programming glitch, the infection was permitted to spread to virtual pets. While the pets were immune and did not die, they served as reservoirs for the pathogen and became a source of immediate reinfection after their owners came back to life or were otherwise cured of the disease.

The programmers scrambled to figure out what was happening as the pandemic raged. Initially, they had no idea why vast numbers of players were suddenly dying. They eventually imposed quarantine measures, isolating infected players from uninfected areas. But this effort failed because players refused to be quarantined, and in any case it was not possible to restrict their movement to the extent required. Ultimately, the programmers resorted to a strategy that doctors and public health officials contending with a real global pandemic do not have: they pulled the plug on the whole world. After the epidemic of corrupted blood had raged unstopped for a week, they rebooted the servers, and the epidemic came to an abrupt and complete halt.

Virtual World, Real Behaviors

These curious events affected literally millions of players, but they also captured the imaginations of people in academia. Microbiologists, mathematicians, psychologists, and epidemiologists were fascinated by the epidemic unleashed by Hakkar. Though the germ and the victims in this outbreak were virtual, the behaviors of the avatars were entirely realistic — so much so that scholars have studied them as indicators of how people might respond to a bioterror attack or the recurrence of a real-world pandemic like influenza.

Some characters in the game had healing powers, and they attempted (largely without success) to cure those afflicted with corrupted blood. They acted altruistically, often rushing to the center of the outbreaks to try to help, and they typically died as a result. Unfortunately, their selfless behavior actually worsened the epidemic in two ways: the healers often became vectors of the infection, and the patients they "cured" remained carriers and went on to infect more people than they otherwise would have if they had simply died. Other characters in the game, lacking the altruism or sense of duty of the healers, fearfully fled infected cities to save themselves but wound up spreading the disease farther. Still others, driven by curiosity or thrill seeking, rushed to the outbreak sites to see what was going on or to see what an infection looked like (victims collapsed in pools of blood). Still others behaved in a sociopathic fashion, deliberately exposing themselves to infection and then quickly transporting themselves to the land of their enemies, or even to their own homeland, to spread the epidemic and cause as many deaths as possible.

Amazingly, a detailed study of the corrupted blood outbreak was published in *Lancet Infectious Diseases,* a medical journal usually devoted to covering the biology and treatment of real-world pathogens.[1] The primary motivation for the study was to see if the virtual world could be used to model real-world behaviors during epidemics. The authors noted that if future virtual epidemics were designed and presented so as to seamlessly integrate within an online game, a reasonable analogue to real-world reactions to epidemics might be studied and even manipulated.

For thousands of years, social interactions were built solely on face-to-face communication. But technology changed this with the invention of ways of broadcasting information (church bells, signal fires, books, bullhorns, radio, television) and ways of communicating person-to-person at a distance (letters, telegrams, phone calls). Today, in addition to the impressive prospect of inhabiting virtual online worlds, we engage in other forms of communication and interaction that have already become plebeian even though they are actually quite remarkable: we text, Twitter, e-mail, blog, instant-message, Google, YouTube, and Facebook one another using technology that did not exist just a few years ago. Even so, there are some things that technology does not change.

The invention of each new method of communication has contributed to a debate stretching back centuries about how technology affects community. Pessimists have expressed the concern that new ways of communicating might weaken traditional ways of relating, leading people to turn away from a full range of in-person interactions with others that, in bygone eras, were necessary and normal parts of life. Optimists argue that such technologies merely augment, extend, and supplement the conventional ways people form connections.

In the case of the Internet in particular, proponents argue that relationships that emerge online can be unfettered by geography and even, perhaps, by awkward constraints attributable to shyness or discrimination. Internet proponents have also seen a benefit to the kind of anonymous and large-scale interactions that are much harder to arrange in the real world. Instead of having personal ties to a small number of people, we have more tenuous ties to hundreds or thousands. Instead of simply knowing who our friends are, and perhaps our friends' friends, we can peer beyond our social horizons and even see graphical depictions of our place in a vast worldwide social network.

Yet, new technologies—whether massively multiplayer online games such as World of Warcraft or Second Life; social-network websites such as Facebook or MySpace; collective information sites like YouTube, Wikipedia, or eBay; or dating sites like Match.com or eHarmony—just realize our ancient propensity to connect to other humans, albeit with electrons flowing through cyberspace rather than conversation drifting through air. While the social networks formed online may be abstract, large, complex, and supermodern, they also reflect universal and fundamental human tendencies that emerged in our prehistoric past when we told stories to one another around campfires in the African savanna. Even astonishing advances in communication technology like the printing press, the telephone, and the Internet do not take us away from this past; they draw us closer to it.

So Real, It's Shocking

To understand how real online behavior can actually be, and how it can be both novel and pedestrian at the same time, we have to go back fifty years to a landmark experiment regarding a rather extreme type of human behavior. Stanley Milgram, famous for the six-degrees-of-separation study and the sidewalk study discussed in chapter 1, did yet another, even more famous, piece of work. Beginning in 1961 at Yale University, shortly after the start of the trial of Nazi war criminal Adolf Eichmann in Jerusalem, and in response to ongoing curiosity after World War II about how people could be induced to "follow orders" to brutalize other human beings, he designed an experiment to evaluate obedience. He wanted to demonstrate how ordinary people would obediently respond to authority and how easily they could be persuaded to inflict pain, even lethal pain, on others.

In Milgram's experiment, the research subjects, confusingly referred to as "teachers," were instructed by an "experimenter" sitting near them to administer electric shocks of increasing voltage to a "learner" whenever the learner gave an incorrect answer to a word-memory problem. Teachers and learners were chosen by lottery. But the lottery was a sham, and the learner was actually a confederate of Milgram's, a forty-seven-year-old Irish-American accountant trained for this role. The experimenter was also a confederate—a rather stern biology teacher wearing a lab coat. The learner was led into a booth, and the teacher—the only person who was not in on what was happening—sat outside, listening to the learner through the wall. As planned, the learner did poorly on the memory test. Milgram's focus was what the teacher could be persuaded to do about it by the experimenter.

In an alarmingly high proportion of cases—65 percent of the time in one experiment—the teachers continued administering "shocks" after wrong answers, at the urging of the experimenter, until the learner reached the maximum, lethal level. In fact, they did so in spite of screams of protest and even banging on the wall by the learner. Even though almost all teachers showed signs of distress at their own behavior, and even though many expressed concern for the learner, they nevertheless continued shocking at the exhortation of the experimenter who sat next to them impassively saying mild things like "the experiment requires that you continue."[2]

Milgram (and others) have replicated this experiment many times, with many variations. For example, Milgram evaluated the extent to which conformity or authority were the motivations for the subjects' behavior. In one experiment, participants were joined in their booth by two additional "teachers" (also members of Milgram's inexhaustible supply of confederates). If these two teachers refused to do what the experimenter instructed, only 10 percent of the subjects gave the maximum shock. Evidently, the presence of others who refused to follow orders made it possible to avoid doing so oneself—a result that was a harbinger of Milgram's later sidewalk experiment. Overall, however, a systematic review in 1999 of many experiments of this kind found remarkably consistent rates of obedience, with 61 to 66 percent of subjects willing to inflict fatal voltages under a diverse set of circumstances.[3]

Milgram proposed two explanations for the obedience he observed. First, people are indeed motivated by conformity. They have a tendency to relinquish their decision making to a group and to its hierarchy, especially when they are under stress. Second, people are able to dissociate from their actions and come to see themselves as instruments of another person's will; hence, they do not consider themselves responsible for their actions.

Milgram's experiment came under extensive criticism as unethical from the moment it was published. Not only did it involve serious deception (the subjects were fooled into thinking that their assignment to their role was by chance and also that the Learner was really suffering); it also involved the infliction of severe distress on the subjects, some of whom thought they had killed another person. Within a few years after Milgram's experiment, in fact, it was impossible to conduct similar experiments.

And this is where the online world enters the scene. In 2006, a group of computer scientists, psychologists, and neuroscientists repeated Milgram's experiment with real teachers but patently virtual learners. These researchers created an immersive environment in which thirty-four subjects were told—and in any case it was plainly obvious—that the learner was a computer animation. Teachers wore virtual-reality headsets and were told to shock the learner with ever-increasing voltage when she did not respond correctly to the word-memory test. The thirty-four subjects were divided into two experimental groups. Twenty-three of them could see and hear the virtual learner as the "pain" was "inflicted," and eleven could communicate with the learner only through a text interface.

In spite of the fact that all participants knew that neither the learner nor the shocks were real, when they could see and hear her they responded as if she were real. When the virtual learner asked for questions to be read louder, the participants responded as if reading louder would indeed make a difference. When the learner complained about the "pain" she was suffering, many participants turned to the experimenter sitting next to them and asked what they should do, whereupon the experimenter would say, "Although you can stop whenever you want, it is best for the experiment that you continue." Twelve of the twenty-three subjects who could see and hear the virtual learner stopped before the end of the experiment, but only one of the eleven who communicated by text did.[4]

These types of experiments offer a possible way out of the ethical proscriptions that arise in conducting such experiments in real life (since neither deception nor real suffering are inflicted). Like dissecting a virtual frog, social scientists can dissect social behavior in a virtual world. But more important for our purposes, these experiments illustrate that life online can both emulate and extend real human interactions. People obey deep-seated rules of human interaction even in these unusual circumstances. The sense of incredible realism that many people experience when interacting in virtual worlds with virtual people is known as *presence*.

My, What a Nice-Looking Avatar You Have

One important way in which virtual worlds differ from the real world is our ability to control our own appearance. In the real world, human manipulation of bodies—whether via clothing or cosmetics, tattoos or plastic surgery—is a cultural universal. Part of the reason for this is that our physical appearance affects the way others treat us. For example, people attract more friends, are paid a higher wage for the same work, and are often seen as more deserving of treatment by health care workers when they are tall or good looking.[5] Our physical appearance, however, also affects how we perceive ourselves and therefore how we act.

Unfortunately, scientists interested in this topic have been hamstrung by the lack of easy ways to transiently and substantially modify someone's looks. Here again avatars provide an alternative. In principle, avatars can have any appearance, and the choices offered by online games are enormous. In the virtual world of Second Life, for example, players can manipulate 150 parameters to change everything from their eye color to their foot size.

It turns out that these seemingly arbitrary manipulations of online appearance affect online interactions as well. In one study, volunteers were assigned avatars who ranged from plain to attractive (regardless of what the volunteers looked like in real life). The volunteers then donned virtual-reality headsets and manipulated their avatars to walk across a room to interact with another avatar controlled by a research assistant who could not see the avatars' virtual faces (and who therefore interacted with all avatars in a uniform fashion). In this way, the investigators were able to cleverly manipulate the avatars so the volunteers could see themselves differently from the way others (in this case, the research assistant) saw them. This is important because if you have an attractive avatar, for example, you might act with more confidence when others treat you with more deference in the online world.

Those volunteers randomly assigned attractive avatars maintained a much smaller interpersonal distance than those with homely avatars: the attractive avatars came within three virtual feet of others, whereas the unattractive avatars came within six feet. Those with attractive avatars showed more self-confidence in other ways as well, such as being more willing to talk about themselves. In other words, volunteers asserted themselves with a confidence befitting how attractive their avatars were, rather than how attractive they were in real life. The investigators called the impact of perceptions of our own appearance on our behavior the *Proteus effect,* after the god in Greek mythology who could assume any appearance he wanted.[6]

In another experiment, people were assigned avatars of different heights. The subjects' avatars then sat down to play the ultimatum game described in chapter 7. Players who had been assigned tall avatars demanded more money during play. When they were given $100 to split with another player, they proposed, on average, that they get $61 and the other player get $39; whereas those with short avatars proposed on average a $52 to $48 split. Conversely, when players were on the receiving end of the game and weighing a $75 to $25 split (an "unfair" split), those with tall avatars accepted it 38 percent of the time, but those with short avatars accepted it 72 percent of the time.[7]

More remarkably, virtual-world interactions can carry over to the real world. After playing the game with randomly assigned avatars, people who had been assigned attractive avatars showed more confidence in the real world. In one experiment, they were shown pictures from an online dating site; volunteers who had been assigned attractive avatars felt more certain that attractive people would be interested in dating them.[8] These kinds of effects even raise therapeutic possibilities. Perhaps using avatars to act out roles (such as being disabled) in virtual environments could increase empathy for those who are disabled. Or, imagine assigning attractive avatars to people with low self-esteem or distorted body images to allow them to experience the world differently.

It's also possible to use such experiments to evaluate the more conventional topic of how our appearance affects how others treat us, and not just how we see ourselves. For example, people appear to follow racial stereotypes online. One study examined the willingness of avatars in the virtual world There.com to help individuals of other races who make a simple request, and it found that requests made by dark-skinned avatars are much less likely to be honored.[9] And avatars obey gender norms consistent with the real world too; for example, pairs of male avatars (regardless of the genders of the real people controlling them) maintain a larger interpersonal distance in the virtual world than do female avatars, and male avatars make less eye contact with other avatars.[10]

Hence, in broad classes of online interactions, despite the digital frontiers we have crossed, we still act in very human ways. We do not leave self-interest, greed, bias, altruism, or affection behind when we cross over to the digital world any more than Hernán Cortés and his men did when they crossed the Atlantic.

Roam Around the World

The online world is but the latest step in a long march of technological and other socioeconomic changes influencing human interactions. Worldwide population growth and urbanization, coupled with astonishing advances in transportation and communication, have had profound effects on whom we meet, do business with, go to school with, and fall in love with.

Over the past two hundred years, the earth's population has grown from less than a billion to nearly seven billion, and more than half of this increase has occurred since 1960. Population density has risen even faster, given the concurrent urbanization that has taken place across the planet. This rise in population density has itself modified the nature of human interactions, as people jostle together in ever tighter spaces. Even more striking, however, has been the change in patterns of human mobility made possible by advances in transportation. While population has gone up sevenfold in two hundred years, mobility has increased over a thousandfold in the same period, further increasing the jostling.

During the late nineteenth century, open-ocean steam navigation eclipsed sail navigation. Especially beginning in the 1860s, fundamental advances in hull, propeller, and engine design dramatically reduced the travel time between far-flung ports. For example, in 1787 it took eight months for the First Fleet to reach Australia from England, a distance of twelve thousand miles over the ocean; but by a century later, the trip took less than fifty days. Not long afterward, in 1925, air travel between the two countries was introduced, and by 1928 daredevil Bert Hinkler made a solo flight in just sixteen days. Amazingly, however, flight times showed a pattern of decline very similar to that of the bygone sea-travel era. Speeds of air travel between the United Kingdom and Australia decreased exponentially to about two days by 1955 and to less than a day now.[11]

The impact of transport technology on the spatial mobility of individuals can also be appreciated by examining data from France over the same two-hundred-year period. During this time, the average distance traveled daily by people taking advantage of ever-newer modes of transport—from horses and barges to railroads and automobiles to bullet trains and airplanes—increased over a thousandfold, from less than a tenth of a kilometer to nearly one hundred kilometers.[12] Coupled with the rise in population in France during the same period, from roughly thirty million to fifty-eight million, all this movement substantially increased the mixing of people.

Changes in human mobility are perhaps best illustrated by some work that epidemiologist David Bradley did while investigating his own genealogy. Bradley documented the travel patterns of his great-grandfather, grandfather, father, and himself during the century leading up to the 1990s.[13] His great-grandfather led a very circumscribed life in a village in Northamptonshire in the British Midlands, never leaving a forty-kilometer-by-forty-kilometer square patch on the surface of our planet. His grandfather moved about a bit more, including traveling to London, but was still limited to a square in southern England with four-hundred-kilometer sides. Bradley's father traveled all over Europe, covering parts of a square with sides measuring four thousand kilometers, and Bradley himself became a globe-trotter, covering the planet's forty-thousand-kilometer circumference. Roughly speaking, with each generation, the range of travel in the Bradley family increased tenfold. To keep this up, Bradley's son would have to be an astronaut.

In an ideal world, in order to truly understand human mobility, we would have some way of implanting people with microchips and tracking them for days on end. This kind of idea struck scientists as totally fanciful, until they realized that people had implanted themselves with such a device. When turned on, mobile phones are constantly connected to a grid of cell towers that can be used, along with some complicated geometry, to track the movements of their bearers. And this is a much more detailed way of understanding human movements than tracking the exchange of dollar bills at WheresGeorge.com.

With this in mind, physicist László Barabási and his colleagues obtained an anonymous sample of over six million mobile phone users' records from one country.[14] Working with one hundred thousand individuals from this data set, they were able to develop mobility records for a six-month period, noting people's position each time they made or received a call, and detecting 16,264,308 changes in location. Barabási found that the overall pattern of human mobility reflected a combination of factors. First, people move in keeping with the Lévy flight pattern (up to a point), as discussed in chapter 5. Second, they vary in how much movement they undertake, with some ranging over much greater distances than others. In part, this complicated picture arises from the fact that we tend to return to the same locations, such as our homes, workplaces, favorite restaurants, and stores, again and again. Moreover, our "flights" are not that random across time; after all, we sleep in the same bed most nights.

This vigorous movement of human beings and the progressive collapse of geographic space have had radical implications for the spread of everything from germs to goods to information to ideas. Today we can form connections over much larger ranges than our hominid ancestors did, and we find a greater variety of individuals with whom to do so for a greater variety of purposes.

Talking at a Distance

In addition to progress in transportation technologies, enormous advances have been made in communication technologies over the same two-hundred-year period. In his book *The Victorian Internet,* journalist Tom Standage documents the ways in which the invention and dissemination of the telegraph affected social life in the nineteenth century.[15] Prior to the invention of telegraphy, people could communicate across distance only as fast as a human could travel to transport a message (on foot, on horseback, or by ship). But the telegraph made a form of communication possible that collapsed both time and distance. This technology was quickly deployed for business and romance, gave rise to new types of interactions in everything from journalism to warfare, and prompted the emergence of new customs and vocabularies.

The invention of the telephone represented a further evolution of this process. Many people were thrilled with the diffusion of the telephone across America in the late nineteenth and early twentieth centuries, and they expected it to break down class barriers and to democratize society.[16] Others thought it would reduce loneliness, especially in agricultural communities, and increase social interactions.[17]

However, presaging contemporary concerns about the Internet, others had a more pessimistic view of the telephone. Many worried that domestic life would be constantly interrupted with demands from the outside, violating the sanctity and tranquility of the home. There were also concerns about loss of privacy due to a nosy, eavesdropping operator, for example. Others thought that rushed telephone conversations would be socially dangerous, as "speakers cannot prepare for or reflect upon the discussion as they can in letters."[18] The telephone also threatened many traditional social customs (like visiting friends without prior notice), and observers worried about the ways in which the telephone would change courtship rituals, leading to inappropriate sexual contact.

Most of these concerns, of course, resemble those expressed about the Internet: it is making interactions rushed; it opens us up to a loss of privacy; it contributes to moral degeneracy. But perhaps more telling than these is the fear that people will replace close friendships within their communities with weak connections to distant friends. Sociologist Charles Horton Cooley observed in 1912 that, because of the telephone and other technologies, "in our own life, the intimacy of the neighborhood has been broken up by the growth of an intricate mesh of wider contacts which leaves us strangers to people who live in the same house...diminishing our economic and spiritual community with our neighbors."[19]

In reality, the telephone did more to expand and strengthen local ties than it did to weaken them. Local interactions were facilitated, and the majority of phone calls were and are to people who live within five miles of a person's home.[20] Most people deny that their telephonic relationships—whether near or far—are shallow. In fact, one early, positive commentator put it quite vividly in 1911. He heralded the development of the telephone by saying: "It has enabled us to be more social and cooperative. It has literally abolished the isolation of the separate family. It has become so truly an organ of the social body that we now enter into contracts by telephone, give evidence, try lawsuits, make speeches, propose marriage, confer degrees, appeal to voters, and do almost everything else that is a matter of speech."[21] The phone supplements, not supplants, social interactions.

O Little Town of Netville

In a repeat of debates about the telephone, discussions concerning the Internet have highlighted how individuals can interact with people they might not otherwise connect with, in distant countries, anonymously, and so on. And there is no doubt that the Internet allows novel connections that were not previously possible. Yet these same technologies also afford opportunities to change local communities and local interactions.

An illuminating illustration is provided by a remarkable series of observations by sociologists Keith Hampton and Barry Wellman. In the late 1990s, they studied a suburb of Toronto, which they called "Netville," that was being equipped with new broadband technology, provided free to residents. Anyone purchasing one of the 109 new, single-family homes in the neighborhood was also supposed to get high-speed Internet access that was always on, a videophone, and online services ranging from a jukebox to health services to local discussion forums. For various reasons unrelated to the residents' preferences, 60 percent did and 40 percent did not get this suite of offerings, thus yielding two groups that could be compared in a kind of natural experiment. Hampton himself lived in Netville from 1997 to 1999 and studied the effect of this technology on community interactions.

Residents who had access to these services developed deeper and broader connections to other residents, with more neighborhood ties. A comparison between wired and unwired residents revealed that the wired residents recognized many more fellow residents by name (twenty-five versus eight residents), talked to twice as many on a regular basis (six versus three residents), visited more of their neighbors at home over a six-month period (five versus three visits), and made many more phone calls to them in a month (twenty-two versus six calls). This electronic communication significantly enhanced, rather than replaced, direct social ties of the kind Americans typically have with their neighbors.

This technology also helped preserve ties and interactions between Netville residents and the friends they had before their move who were located at some distance from Netville. For example, compared to the nonwired residents, the wired residents had a slight increase in their contact with members of their social networks who lived more than fifty kilometers away, thus counteracting the possibly adverse impact of moving on the maintenance of social ties.

These social ties were in turn put to collective use to mobilize community activities and events. In an ironic twist, one of the targets of the residents' mobilization was the developer of Netville who had installed the technology. Netville residents used their connections, online and offline, to work together to protest defects in the construction of their homes and to pressure the developer to fix them. The rapidity with which residents were able to coordinate their efforts caught the developer totally by surprise and obliged him to address residents' concerns "with more resources and with greater speed than he anticipated."[22] The residents also coordinated a drive to get town officials to deny the developer a permit to work on a second housing development. As Hampton wryly summarized: "Based on his experiences in Netville, the developer acknowledged that he would never build another wired neighborhood."[23]

From Six Degrees to Facebook

Online social-network sites, which have become very popular in the past few years, are services that allow users to construct a public or semipublic personal profile in a limited-access environment, display a list of other users with whom one shares a connection, and view and navigate one's own connections and the connections of others within the system.[24] While it is possible to make new friends and connections in these sites, this is not their primary purpose (unlike dating sites such as Match.com). Some sites support existing friend-ships, like Friendster.com, and others serve the needs of particular ethnic, political, religious, or professional audiences, like the net-work of scientists and teachers at MySDScience.com. Sites also vary in their rules regarding privacy, who can join, what can be posted, and how network connections are seen and traversed by others. The distinctive feature of social-network sites is that it makes our web of connections visible to the user and to others. Moreover, unlike other sorts of online groups or communities like wikis and listservs, social-network sites are organized around people, not topics.

Hundreds of millions of people have integrated the use of social-network sites into their daily lives. They get status updates on their friends, make new connections, play games, and post favorite links online every day. But at their core, social-network sites primarily reflect offline interactions. Although they allow us to maintain con-tact with people to whom we would otherwise be tied only weakly (such as former roommates, high school classmates, people we meet briefly at parties, and so on), they are not organized around the introduction of strangers.

The first recognizable online social-network website, SixDegrees .com, was launched in 1997.[25] It attracted many users but failed as a business in 2000, probably because the market was not yet ripe for the concept. Part of the problem was that, like the telephone or the fax machine, an online social network is not useful until many other people are also using it.

In 2002, Friendster was launched to compete with Match.com. Unlike Match.com and similar dating sites that were focused on facilitating the introduction of strangers, Friendster exploited the idea that friends of friends would be a better pool from which to draw romantic partners. In essence, it was the computer-facilitated version of the real-world process of mate search in social networks that we discussed in chapter 3. Friendster grew rapidly, acquiring three hundred thousand users by 2003. But when it started getting media attention, interest in the site skyrocketed, and it encountered technical and social problems. Technically, its servers and databases were not up to the task of meeting the needs of a network that was growing exponentially in its complexity. Socially, as large numbers of people flocked to the site, they changed its culture, rather like a summer influx of rowdy, drunk college students to a sleepy Greek island, leading to a breakdown in traditional norms and cohesion among prior users.

For example, at its inception, Friendster limited the profiles one could see to four degrees of separation (friends of friends of friends of friends). Interestingly, this practice was just one degree beyond the normal sphere of influence (recall the Three Degrees of Influence Rule) and the train of introductions that one has access to in the real world (as we saw in chapter 3). That is, implicitly, the design of Friendster seemed to leverage computer technology to broaden our social horizons by one degree. But the new users tried to push even farther into the network by creating shortcuts. The strategy was to befriend a few strangers at four degrees of separation in order to gain access to distant parts of the social network that could not previously be seen. However, the ability to see beyond the natural social horizon meant that a much larger percentage of the friendships on the site were not based on real-world connections. Partly as a consequence, Friendster faded in popularity in the United States.

MySpace was launched in 2003, and it capitalized on the disaffection of former Friendster users. From the beginning, it placed a strong emphasis on attracting fans of indie rock bands, and it allowed users to develop connections to the bands and to other fans of the bands. MySpace also allowed users to develop highly personalized profiles and to cut and paste material into their profiles from elsewhere. As a general social networking site, however, it would soon be overshadowed by a competitor.

The online social network Facebook began at Harvard in 2004, but its story is actually rooted in a real-world phenomenon. The name comes from a long-standing pre-Internet institution at Harvard: each year the college would publish and distribute a book that showed all the students in a given class and where they lived on campus. It was like a telephone directory with photos, and students came to depend on it for social life. One year, when production of the facebook was delayed due to publishing errors, four students in one of the Harvard dormitories actually went on a hunger strike.[26]

In a practice that anticipated the online version, some students would literally "shop" for dates using the facebook, while a handful of more ambitious students would attempt to memorize the names and faces of every single person listed. The earliest reference to the facebook was written in the *Harvard Crimson* in 1979 by a young Susan Faludi (who would later win a Pulitzer Prize for Explanatory Journalism). She reported that it was being used at the time to evaluate potential mentors for first-year students: "We used the facebook to see what people were like....Sometimes you can tell from a picture."[27]

Twenty-five years later, Mark Zuckerberg, who was a sophomore at Harvard at the time, took the facebook online, and it became so popular that it quickly spread to other institutions. Originally, users had to be members of a college community, and the site fostered a sense of intimacy and privacy—the online realization of a sheltered offline world. Members of the community could see anyone's profile within the community—just as if they had bumped into them on campus, only now with anonymity. Moreover, and crucially, the online links that could be formed were visible to others. Within a year, membership opened up to high schools as well, and then later to geographically specified communities and corporate networks.

In June 2008, Facebook surpassed MySpace in total worldwide users to become the largest online social network, and, as of early 2009 over 175 million people were registered and actively using the site.[28] One feature that has probably helped Facebook succeed where Friendster failed is the restriction on whom users can see in the network. Unlike Friendster, which showed people up to four degrees away, Facebook only allows people to see direct friends (one degree) and occasionally friends of friends (two degrees, via the "People You May Know" feature). This reduces the number of links between total strangers and makes people feel as though their online life is relevant to their real-world social networks.

Whether social-network sites will endure and in what form is unclear, but after a decade of innovation, it seems that such sites, in some form at least, are here to stay. Social-network sites are now adding real-time communications features, such as instant messaging, e-mailing, and so forth. At the same time, other sites with user-generated content (such as Flickr.com for photos and iLike.com for music) are adding social-network features. Even older adults are joining online social-network sites like Eons, ReZoom, and Multiply. Slowly but surely, we are taking our real lives online.

Massive and Passive

Second Life, World of Warcraft, Facebook, and MySpace are entirely separate sites. To participate in all four, one must maintain a profile and an identity separately at each site. However, it will soon be possible to have a single identity managed from a single profile that allows one to traverse many virtual worlds and social networks. Something similar happened with e-mail. The original e-mail programs only allowed people to e-mail others who had the same program. That restriction was soon dropped and interoperability became standard as people demanded one entry point to all e-mail networks. We may actually move away from sites that are designed exclusively for social networking. Certainly, these gardens will become less walled off, and the ability to communicate with people in multiple social-network sites, transport friend lists from one site to another, or open up the content to searches from elsewhere will increase. But, in addition, other sources of long-used data may become the basis for social networking online.

People's e-mail lists already capture much valuable social-network information, and they do so dynamically and in many ways more effectively than explicit social-network sites. Address books and calendars also provide valuable information. A person's e-mail in-box and out-box indicate who one is in touch with, when, and how often. These data could be used to draw networks and to order people in terms of frequency or recency of contact. E-mail even allows one to grade the directionality of a tie: you get a lot of mail from Tom and Harry, but you only respond to Tom. Thus, e-mail systems could provide the infrastructure for social networking even though they are not explicitly designed for this purpose.

Online social networks currently require us to provide explicit information about our connections to others and about our daily activities, but soon such networks will be implicit. New systems are evolving that automatically generate massive amounts of passively collected data online and allow us to automatically track our friends. With iLike you can opt to show friends what songs you are listening to on your computer or your iPhone. You can also have your calendar automatically published on Google and even continuously broadcast your GPS position via your mobile phone with applications like Twinkle. These applications will soon give us access to even more detailed information about our friends, with algorithms that use passive data to make inferences about our friends' lives. A flurry of e-mails to a new person might suggest our friend made a new friend. Repeated visits to a new website might indicate our friend has a new hobby. In fact, corporations are already rushing to monetize these new technologies because they can help to focus advertising and predict what products and services a person will buy. If I know what your friends are doing, I can make a good guess as to what you will soon be doing.

The recent surge in mobile phones, the Internet, and social-network sites has shifted our ability to stay in touch with one another into overdrive, causing us to become hyperconnected. This new technology can give us a sense of how connected or unconnected we are in real time. Might this technology help us to increase the efficiency of language just as language represented an advance over grooming? In what ways might technology change social networks?

The Internet makes possible new social forms that are radical modifications of existing types of social-network interactions in four ways:

1. *Enormity:* a vast increase in the scale of our networks and the numbers of people who might be reached to join them
2. *Communality:* a broadening of the scale by which we can share information and contribute to collective efforts
3. *Specificity:* an impressive increase in the particularity of the ties we can form
4. *Virtuality:* the ability to assume virtual identities

Too Many Friends?

While many users of social networking sites have hundreds or even thousands of people they list as friends, it turns out that the average user has approximately 110 friends on Facebook.[29] And it is clear that only a subset of these are close friends. To figure out who was close and who was not, we developed a "picture friends" method based on the photographs that people post on their Facebook pages. The idea is that two people who post and "tag" pictures of each other are much more likely to be socially close than those who do not. We studied all the Facebook pages at a college (we can't say which one), and when we counted the number of picture friends that students had, we found that, on average, just 6.6 were close friends.

Amazingly, these observations show how closely online social networks resemble offline networks. The total number of friends that people have online, on average, is not far from 150, Dunbar's number (discussed in chapter 7). And the number of close friends is not far from the core network size of four (discussed in chapter 1). Online networks therefore do not appear to expand the number of people with whom we feel truly close, nor do they necessarily enhance our relationships within our core groups. We are still bound by our primate tendencies and abilities.

Yet, social-network sites do offer new opportunities. A group of Facebook "friends" is very different from a group of people inhabiting a Paleolithic village, not so much in terms of who we are but, rather, what passes for coherent or normal social interaction. Social-network sites can expand and redefine what counts as a friend, while at the same time facilitating the maintenance of ties within this broader group of people. Social-network sites are used to keep tabs on real friends and relatives, of course, but most people have online connections to others whose phone numbers, for example, they might not have, whom they might not be able to recognize on the street, and, frankly, whom they might not feel comfortable chatting with in a bar.

The friends we have in our online social networks differ from offline connections in other ways: such friendships tend to be cumulative (people tend to add connections online and not cut them), and the nature of the interaction is strongly influenced by the medium (briefer bursts of activity rather than more sustained conversations, for example). In online networks, moreover, we not only manage our direct relationship to all these people; we also monitor all of their relationships with one another to a much greater degree than we would in the offline world. Every breakup between our friends is reported with a little red broken heart next to a friend's name; in online networks of high school and college students, the average news feed probably contains dozens of souls in need of consolation. Suddenly, we are much more aware of the everyday lives of people we might have forgotten or lost touch with in our face-to-face social networks.

The relevance of online ties can be seen graphically. Plate 8 illustrates the difference in connectedness between real-life and online networks for 140 university students from our Facebook study. First we show the web of close friendships, based on our picture-friends algorithm. Then we add ties based on belonging to the same club (top right) or being roommates (bottom left). Finally, in the lower right corner, we add the Facebook friendships. What was once a gossamer web became a tangled ball of yarn! And this is just part of the network. When we look at all 1,700 students in the class at this university, the interconnections expressed on Facebook become impenetrably dense.

Reality and Wikiality

The massive scale of online interactions has made possible a wide variety of acquaintance networks that never before existed. The result is a huge increase in the sharing of information. There are online vacation planning sites where people post photos of destinations for other travelers, collaborative efforts to correct online geography databases (people reaching a dead end in a road not shown on a map promptly and electronically notify the map provider), and even sites like CouchSurfing.com that provide lists for numerous cities where members can sign up to sleep on a stranger's couch if they need a place to stay while traveling. Open source software sites like SourceForge.net allow people to contribute code to improve a computer program, leading to the development of web browsers like Firefox and operating systems like Linux that compete head-to-head with products from Microsoft and Apple.

But the most widespread example of this new form of social interaction is the wiki. Taken from the Hawaiian word for "fast," wikis are designed to allow anyone with access to the wiki to make changes to its content. This allows groups of people who care about the same information to pool their resources and work together. By making it almost costless to connect, wikis harness the power of a million tiny acts of kindness to create something new and powerful. The best-known example of a wiki is Wikipedia, an online encyclopedia that at last count contained more than twelve million entries written in two hundred languages. There is no strong centralized authority on Wikipedia. Like any other wiki, it is maintained by volunteers who collaborate with one another and make their own rules about how to interact.

Comedy Central's Stephen Colbert pokes fun at these new forms of online collaboration in his show *The Colbert Report*. As a parody of personality-based news-show hosts like Bill O'Reilly and Rush Limbaugh, Colbert plays on the idea that influential hosts can get members of their audience to believe absolutely anything. And if you let these people interact online, they will create their own reality that has no basis in the real world. Colbert defines this "wikiality" as "a reality where, if enough people agree with a notion, it becomes the truth."[30] To demonstrate his capacity to influence wikiality, he once joked on *The Colbert Report* that the world population of elephants was no longer in need of protection because it had tripled in the past decade. Within minutes, the Wikipedia entry for elephant had been changed to highlight this "fact."[31] But moments later the entry was corrected by volunteers who regularly contributed to it. A struggle ensued, and the pranksters eventually had to yield. People with a commitment to preserving correct information won the day, and the entry returned to its original status.

If you think that the uproar caused by a joke was a struggle, consider what happened to the entry for Sarah Palin when she was nominated as the Republican party's candidate for vice president. As one of the most polarizing national candidates in modern American politics, she inspired an "edit war" that generated hundreds of changes per day to her Wikipedia entry during the campaign. Supporters deleted entries referring to her original support for the "Bridge to Nowhere" boondoggle, and detractors added rumors that she was faking her pregnancy to cover up her sixteen-year-old daughter's alleged pregnancy. Yet in spite of truth-distorting edits like these from extremists on both sides, the overall information about Palin remained relatively free of bias. Wikipedians strongly committed to their self-created POV rule (that entries should not be biased by a particular person's point of view) vigilantly monitored all changes to ensure that they were relevant and supported by published sources. Any change that showed too much bias was quickly removed. And countless edits that were in the gray area between bias and fact elicited debates on the "talk" page for Palin's entry, as contributors struggled to decide what did and did not constitute factual information.

The success of sites like Wikipedia is counterintuitive. As wiki expert Anja Ebersbach wrote: "Most people, when they first learn about the wiki concept, assume that a website that can be edited by anybody would soon be rendered useless by destructive input. It sounds like offering free spray cans next to a gray concrete wall. The only likely outcome would be ugly graffiti and simple tagging, and many artistic efforts would not be long lived. Still, it seems to work very well."[32] In fact, a study published in the scientific journal *Nature* revealed that a typical article in Wikipedia was almost as accurate as a typical article in the *Encyclopedia Britannica*.[33]

Wikis are analogous to elections, markets, and riots but concern knowledge instead of votes, transactions, and emotions. Like guessing the weight of an ox, individuals work independently but collaborate to create something not present in the individuals and not within the reach of any individual. By averaging and collecting information from multiple sources, wikis create a path to knowledge akin to flocks of birds naturally choosing which way to fly.

The reason Wikipedia works so surprisingly well is that social networks emerge around each topic. These networks contain *cooperators* (people who contribute new and unbiased information) and *free riders* (people who want to use the credibility of the information established by others for their own purposes). If those were the only types, it might be easy to believe that Wikipedia would fail. But a third type also exists: the *punisher.* Thousands of vigilantes patrol Wikipedia, reverting malicious edits and leaving personal notes on the "talk" pages of the perpetrators. They can even band together to prevent certain users from making further changes. Hence, amazingly, what we see online is what might have been happening at the dawn of human civilization, just as discussed in chapter 7. We do not cooperate with one another because a state or a central authority forces us to. Instead, our ability to get along emerges spontaneously from the decentralized actions of people who form groups with connected fates and a common purpose.

Finding a Needle in a Haystack

As we saw in chapter 3, millions of people today are using the Internet to connect to romantic partners. Although the search for true love is not new, the Internet makes it much easier to find and interact with many more potential mates, with a tremendous degree of specificity. But it is now also easier to find other sorts of people in other domains of our lives.

In the fall of 2005, Allison Pollock was a fifteen-year-old girl with an unusual condition known as laryngeal cleft, which created a passage between her larynx and esophagus. As a result, food and liquids often wound up in her lungs, causing recurrent bouts of pneumonia. After a long medical course, she came to Children's Hospital in Boston and had specialized surgery that cured her problem. An article about her appeared in the hospital's online magazine, and a young man by the name of Sam Kase who had the same condition read it and decided to track her down: "Allison had the surgery at about the same age as I was, when she was fifteen, back in 2005," he said, "I assumed she was either now a senior in high school or freshman in college. So chances were good that she'd have a profile on Facebook. I hoped that if I found her, I could ask a couple of questions about the surgery and her recovery." Allison, altruistically and unsurprisingly, was happy to respond to this contact, and she wound up sharing frequent e-mails with Sam both before and after his surgery, and, eventually, the two met in person, along with their parents.[34]

Some websites are set up explicitly for such purposes. For example, the welcoming material of a website for the Association of Cancer Online Resources notes: "You are not alone. Use ACOR mailing lists to connect with people like you online and share support and information.... ACOR is one of the original social networks.... Since September 1995, ACOR users interested in specific and often rare forms of cancer have benefited from the collective intelligence of hundreds of patients and caregivers."[35] In 2008, this website delivered more than 1.5 million e-mails worldwide per week.

All this connectedness might sound like a good thing. After all, who would not want to be able to find exactly the person one is looking for? But connectedness also comes with a price. Being more connected means we can find more people, but it also means more people can find us. Not all of those people are well intentioned, and not all connections are positive.

For example, some people use the Internet primarily to increase the number of different people they can have sex with. In turn, this can increase the risk of sexually transmitted diseases, and there have been a number of outbreaks related to Internet use. In a paper entitled "Tracing a Syphilis Outbreak Through Cyberspace," a set of public health physicians in San Francisco reported that if you found your partner online, you were more than three times more likely to get an STD from them than if you found your partner the old-fashioned way.[36]

Unfortunately, those most susceptible to the dark side of online social networks are the young, and the Internet is the new teen hangout. More than 80 percent of American teenagers use the Internet, and nearly half do so every day.[37] Well over 75 percent use e-mail, instant messaging, or other online communication technologies. More than 50 percent have several e-mail addresses or screen names through which they can interact anonymously with others in chat rooms, online forums, and other venues.[38] Online interactions provide valuable social support for potentially isolated adolescents, but they also can routinize and even validate dangerous behaviors such as anorexia, vandalism, and suicide.

Moreover, the specificity, breadth, and immediacy of online social-network culture, with everything from text messages to social-network sites, makes it much more likely that such behaviors will spread. Online, instantaneous feedback and reinforcement can be provided for a brief odd or negative thought or impulse, which might have dissipated on its own in previous generations that lacked such communication technology. Of course, teenagers have always influenced one another, but in the past it typically took much more effort to get that sort of reinforcement. Now it's just a button away.

For example, one study found more than four hundred message boards on the Internet devoted to "cutting."[39] Cutting has been called the "next teen disorder," and it refers to a variety of self-injurious behaviors whose prevalence is probably in excess of 4 percent of teenagers. The behavior has been increasing dramatically in recent years, and many clinicians suspect social contagion as an explanation because cutting follows epidemic-type patterns, with outbreaks in institutions, for instance.[40] For complicated physiological and psychological reasons, teenagers engaged in cutting report that they do so in order to "relieve their distress."

A comprehensive study of message-board postings about this behavior revealed that the most common kind of posting, 28 percent of messages, provided support to others, with comments such as, "We're glad you've come here," or "Just relax and try to breathe deeply and slowly."[41] Alas, 9 percent of posts were about the perceived addictiveness of cutting ("I may try and quit, but even if I succeed, I'll always dream of razor blades and blood"), and 6 percent of posts are about cutting techniques.

Another set of people who have congregated online are those who experience paranoid delusions. For example, there is a group called Freedom from Covert Harassment and Surveillance where several hundred regular users discuss their ideas about being watched. "It was a big relief to find the community," said Derrick Robinson, fifty-five, a janitor in Cincinnati and president of the organization. "I felt that maybe there were others, but I wasn't real sure until I did find this community." Another group is devoted to "gang stalking," the belief that one is subject to "a systemic form of control, which seeks to destroy every aspect of a Targeted Individual's life. The target is followed around and placed under surveillance by Civilian Spies/Snitches 24/7."[42]

These sites provide to delusional individuals the potent, reassuring, and calming experience that we all crave: being understood by others. For once, these people can find many others who provide reassurance that they are not crazy. The ability to connect to others online could therefore be helpful socially, providing a degree of support and human contact in regular, daily life that might not otherwise be possible. But that support may make things worse for them psychologically. "The views of these belief systems are like a shark that has to be constantly fed," notes Dr. Ralph Hoffman, a psychiatrist from Yale. "If you don't feed the delusion, sooner or later it will die out or diminish on its own accord. The key thing is that it needs to be repetitively reinforced."[43] The Internet, alas in this case, affords just this opportunity.

A Whole New You

While some people take their delusions online, others use the Internet to leave their real experiences behind. In virtual worlds, it is possible for people to have a "second life" and interact without real-world constraints. Physically disabled people might have able-bodied avatars, or men might pretend to be women and experiment with social roles in a way that was simply not possible before the Internet existed. These are indeed novel social forms and not merely modifications of existing types of social-network interactions.

And these new forms can blur the lines between real and virtual worlds. In one online game, a forty-three-year-old Japanese woman was married to a thirty-three-year-old office worker whom she did not know personally. The game was proceeding normally when suddenly he divorced her without warning. Although the marriage was virtual and purely imaginary, she became so angry that she used information she had about him to kill his avatar. She had not plotted any revenge in the real world; still, she was arrested by real police officers and later faced a real penalty of up to five years in prison and a real $5,000 fine for her destructive actions online.[44]

Even weirder was the following sequence of events. Amy Taylor, then twenty-three, met her husband, David Pollard, then thirty-five, in an Internet chat room in 2003. They were married in real life in 2005 but had an extremely lavish, parallel wedding in Second Life. After their wedding, Taylor caught her husband's avatar having virtual sex with the avatar of another woman playing the part of a prostitute. She had been suspicious for a while, and, in a surreal twist, had hired a virtual detective agency to track his online activities. "He never did anything in real life," she admitted, "but I had my suspicions about what he was doing in Second Life." In her (real) divorce filings, Taylor described her husband's activities as "committing adultery." Pollard admitted to having had an online relationship but denied any (real) wrongdoing. For her part, Taylor was reported to be in a new relationship with a man she subsequently met playing World of Warcraft.[45]

How is Mr. Pollard's activity different from, say, finding your spouse with pornography? It may well be the connection. He was not just looking at a naked person on the computer screen, nor even a naked avatar, but he was establishing a connection. Or so it must have seemed to Ms. Taylor, and this was the key fact for her.

Still, if it is the case that people see themselves differently online—that people with attractive avatars act more gregariously or behave more beneficently—then it may be the case that online communities will come to have features not seen in real communities, features we have not yet experienced or imagined. Our virtual worlds may seem better than our real world, not just because of what the programmers build in to them, but because of the way we, as human beings, naturally come to behave in these new environments.

The Same but Different

We may use the Internet to find people we already know in the offline world, bringing these relationships online. We may use the Internet to meet new people online with the hope of forming connections in the real world. Or our connections can start and remain in their respective online or offline worlds. In many ways, our online connections resemble our offline ones, but in other ways, our online connections represent entirely new ways and patterns of interacting. The hyperconnectedness made possible by the online world exploits an ancient biological machinery in new ways but still in the service of an ancient goal.

Online networks provide new avenues for influence and social contagion. The rapid organization facilitated by online interactions in Netville, the way that impulses as diverse as racism and altruism are manifest online, and the way the Obama campaign and the Colombian activists used the Internet to mobilize supporters all suggest that social influence can spread through the Internet as it does in real-world social networks.

But some things might spread more easily than others. As we saw in chapter 2, the spread of emotions seems to require face-to-face interaction. So while online connections increase the frequency of contact, it is not clear whether this has the same effect as being present in person. In contrast, in chapter 4 we showed that frequency of contact is not as important for the spread of social norms. The eating, drinking, and smoking habits of our friends who live hundreds of miles away appear to have as much influence as the habits of our friends who live next door. This means that ideas about behavior can spread even in the absence of frequent direct personal contact. Yet the spread of these ideas appears to rely on deep social connections; hence, additional, weak contacts online may have little or no effect on our susceptibility to changing norms. Overall, the evidence from real-world networks suggests that online networks can be used to enhance what flows between real-world friends and family, but we do not yet know whether the Internet will increase the speed or scope of social contagions in general.

Our interactions, fostered and supported by new technologies, but existing even without them, create new social phenomena that transcend individual experience by enriching and enlarging it, and this has significant implications for the collective good. Networks help make the whole of humanity much greater than the sum of its parts, and the invention of new ways to connect promises to increase our power to achieve what nature has foreordained.

CHAPTER 9

The Whole Is Great

B abylon, the first city to be built after the great mythical flood, was one in which, the book of Genesis tells us, humanity was united: "And the Lord said, 'Behold, the people is one, and they have all one language...and now nothing will be restrained from them, which they have imagined to do.'"[1] And the first thing that the unrestrained residents of Babylon imagined to do was to build a tower so immense that it would reach the heavens. Genesis recounts that God punished the people by destroying the tower, giving them multiple languages, and scattering them across the earth. This story illustrates the folly of hubris, and we usually focus on the polyglot consequences. Often overlooked, however, is the fact that the Babylonians were punished not so much by being given different languages but, rather, by becoming disconnected from one another.

By banding together, the citizens had been able to do something— build the tower—that they could not have done alone. Other stories from the Bible allude to the power of connections but put a more positive spin on what connected humans can do. When Joshua and the Israelites arrived at the gates of Jericho, they found that the walls of the city were too steep for any one person to climb or destroy. And then, the story goes, God told them to stand together and march around the city. When they heard the sound of the ram's horn, they "spoke with one voice"—in a kind of synchronization like *La Ola*— and the walls of Jericho came tumbling down.

Observations about connection and its implications are ancient, in no small part because theologians and philosophers, like modern biologists and social scientists, have always known that social connections are key to our humanity—full of both promise and danger. Connections were often seen as what distinguished us from animals or an uncivilized state.

In 1651, the English philosopher Thomas Hobbes engaged in a thought experiment in which he described the prototypic condition of human existence. In a "state of nature," he supposed in his famous work *The Leviathan*, there is *bellum omnium contra omnes*, a "war of all against all." There is anarchy. It is, in fact, Hobbes who observed that the "life of man [is] solitary, poor, nasty, brutish, and short."[2] Hobbes's use of *solitary*—which is often, unaccountably, clipped from the phrase—suggests that a disconnected life is full of woe.

Given these grim circumstances, Hobbes theorized, people would have chosen to enter into a "social contract," sacrificing some of their liberty in exchange for safety. At the core of a civilized society, he argued, people would form connections with one another. These connections would help curb violence and be a source of comfort, peace, and order. People would cease to be loners and become cooperators. A century later, the French philosopher Jean-Jacques Rousseau advanced similar arguments, contending in *The Social Contract* that the state of nature was indeed brutish, devoid of morals or laws, and full of competition and aggression. It was a desire for safety from the threats of others that encouraged people to band together to form a collective presence.

This progression of human beings out of such an ostensibly anarchical condition into ever larger and ever more ordered aggregations—of bands, villages, cities, and states—can in fact be understood as a gradual rise in the size and complexity of social networks. And today this process is continuing to unfold as we become hyperconnected.

The Human Superorganism

The networks we create have lives of their own. They grow, change, reproduce, survive, and die. Things flow and move within them. A social network is a kind of human superorganism, with an anatomy and a physiology—a structure and a function—of its own. From bucket brigades to blogospheres, the human superorganism does what no person could do alone. Our local contributions to the human social network have global consequences that touch the lives of thousands every day and help us to achieve much more than the building of towers or the destruction of walls.

A colony of ants is the prototypic superorganism, with properties not apparent in the ants themselves, properties that arise from the interactions and cooperation of the ants.[3] By joining together, ants create something that transcends the individual: complex anthills spring up like miniature towers of Babylon, tempting wanton children to action. The single ant that finds its way to a sugar bowl far from its nest is like an astronaut stepping foot on the moon: both achievements are made possible by the coordinated efforts and communication of many individuals. Yet, in a way, these solitary individuals—ant and astronaut, both parts of a superorganism—are no different from the tentacle of an octopus sent out to probe a hidden crevice.

In fact, cells within multicellular organisms can be understood in much the same way. Working together, cells generate a higher form of life that is entirely different from the internal workings of a single cell. For example, our digestion is not a function of any one cell or even one type of cells. Likewise, our thoughts are not located in a given neuron; they arise from the pattern of connections between neurons. Whether cells, ants, or humans, new properties of a group can emerge from the interactions of individuals. And cooperative interactions are hallmarks of most major evolutionary leaps that have occurred since the origin of life—consider the incorporation of mitochondria into eukaryotic cells, the agglomeration of single-cell organisms into multicellular organisms, and the assembly of individuals into superorganisms.[4]

Social networks can manifest a kind of intelligence that augments or complements individual intelligence, the way an ant colony is "intelligent" even if individual ants are not, or the way flocks of birds determine where to fly by combining the desires of each bird.[5] Social networks can capture and contain information that is transmitted across people and time (like norms of trust, traditions of reciprocity, oral histories, or online wikis) and can perform computations that aggregate millions of decisions (such as setting a market price for a product or choosing the best candidate in an election). And networks can have this effect regardless of the intelligence of the individual members. Consider, for example, that the way humans laid a rail network throughout England in the twentieth century resembles the way fungi (another species that forms superorganisms) collaboratively explore a patch of ground in the forest in order to exploit and transport resources by creating a network of tubes.[6] Fungi can even "collaborate" to find the best path through mazes into which they have been placed by human experimenters.[7]

Social networks also have a memory of their own structure (staying intact even if people come and go) and their own function (preserving a culture even when people come and go). For example, if you join a trusting network of people, you benefit from that trust and are shaped by it. In many cases, it is not just that the people in your network are more trusting, or even that their trusting behavior engenders trust in you; rather, the network facilitates this trust and changes the way individuals behave.

Like living creatures, networks can be self-replicating. They can reproduce themselves across space and time. But unlike corporeal organisms, networks can, if disassembled, reassemble themselves at a distance. If every person has a memory of whom he or she is connected to, we can cut the connections and move all the people from one place to another, and the network will reappear. Knowledge of one's own social ties means that the network can reemerge even though no single person knows how everyone else in the network is connected.

Networks are also self-replicating in the sense that they outlast their members: the network can endure even if the people within it change, just as cells replace themselves in our skin, computers are swapped out on a server farm, and new buyers and sellers come to a market that has been located in the same place for centuries. In one study of a network of four million people connected by their phone calls, researchers showed that, paradoxically, groups with more than fifteen interconnected people that experienced the greatest turnover endured the longest.[8] Large social networks may in fact require such turnover to survive, just as cell renewal is required for our bodies to survive.

These observations highlight another amazing, organism-like property: social networks are often self-annealing. They can close up around their gaps, in the same way that the edges of a wound come together. One person might step out of a bucket brigade, but then the two people he was connected to will move closer to each other, forming a new connection to fill in the gap. As a result, water will continue to flow. In more complicated, real-life networks, it seems likely that the very purpose of redundant networks ties, and of transitivity, is precisely to make the networks tolerant of this kind of loss, as if human social networks were designed to last.

Like a worldwide nervous system, our networks allow us to send and receive messages to nearly every other person on the planet. As we become more hyperconnected, information circulates more efficiently, we interact more easily, and we manage more and different kinds of social connections every day. All of these changes make us, *Homo dictyous,* even more like a superorganism that acts with a common purpose. The ability of networks to create and sustain our collective goals continues to strengthen. And everything that now spreads from person to person will soon spread farther and faster, prompting new features to emerge as the scale of interactions increases.

It's Neither Yours Nor Mine

The social networks we create are a valuable, shared resource. Social networks confer benefits. Alas, not all people are in the best position to capture these benefits, and this raises fundamental questions of justice and public policy.

Social scientists call such a shared resource a *public good.* A *private good* is one the owner can exclude others from consuming, and one that, once consumed, cannot be consumed again. If I own a cake, I can prevent anyone else from eating it, and once I eat it myself, there is none left for anyone else. A public good, in contrast, can be consumed without harming the interests of others, and without reducing others' ability to use it. Think of a lighthouse. One ship making use of the light to avoid colliding with the rocks does not prevent another ship from doing the same. Public radio, Fourth of July fireworks, and municipal water fluoridation are other examples of public goods. Not all public goods are man-made, of course. Think of the air. One person breathing does not make anyone else have less air, nor does it prevent anyone else from breathing.

Other public goods are less tangible even than light or air. Think of civic duty. As Alexis de Tocqueville argued in the early nineteenth century, if everyone feels the obligation to maintain a civil society, to act in a trustworthy way, and to volunteer for the nation in times of attack, then all citizens can benefit from these traditions and norms. And the benefit to one person does not reduce the benefit for others.

But public goods are difficult to create and maintain. It often seems that no one has an incentive to care for them, as a breath of not-so-fresh air in any polluted city demonstrates. Hence, public goods often arise as by-products of the actions of individuals acting with some self-interest. A shipping company or port authority that builds a lighthouse to safeguard its own ships ends up safeguarding all ships.

Some public goods get better the more they are produced. A classic example of a particular kind of network good is a telephone or fax machine. The first person to get a fax machine finds that it is worthless because there is no one to fax anything to. However, as more and more people acquire fax machines, they become more and more valuable. A similar—if more abstract—example of a network public good is trust. As discussed in chapter 7, trust is most valuable when others are also trusting; and being trusting in a world of free riders is very painful. But many other human behaviors and beliefs increase in value in this way. For example, the positive effects of religiosity on well-being is higher in countries where average religiosity is higher.[9] Like fax machines, religion is more useful if others also believe, in part because religion works to enhance well-being by fostering social ties.

The social networks that humans create are themselves public goods. Everyone chooses their own friends, but in the process an endlessly complex social network is created, and the network can become a resource that no one person controls but that all benefit from. From the point of view of each person in the network, there is no way to tell exactly what kind of world we inhabit, even though we help create it. We can see our own friends, family, neighbors, and coworkers, and perhaps we know a little bit about how they are tied to each other, but how we are connected to the network beyond our immediate social horizon is usually a mystery. Yet, as we have seen time and again, the precise structure of the network around us and the precise nature of the things flowing through it affect us all. We are like people on a crowded dance floor: we know that there are ten people pressed up against us, but we are not sure if we are in the middle or at the edge of the room or whether a wave of ecstasy or fear is spreading toward us.

Of course, not all networks create something that is useful, valuable, and shared, let alone something that is positive. When we say "good" we really just mean any old thing: pistols and poisons are goods too. And networks can function as conduits for pathogens or panic. Indeed, social networks can be exploited for bad ends. As we noted in chapter 1, violence spreads in networks, as does suicide, anger, fraud, fascism, and even accusations of witchcraft.

The interpersonal spread of criminal behavior is an illuminating example of a bad network outcome. One persistent mystery about crime is its variation across time (fluctuating from year to year) and space (varying in adjoining police precincts or jurisdictions). For example, in Ridgewood Village, New Jersey, there are 0.008 serious crimes per capita, whereas in nearby Atlantic City, there are 0.384—a nearly fiftyfold difference. This variation seems too great to be explained by some kind of disparity in the costs and benefits of crime, or even in observable features of the environment or the residents, such as the availability of after-school programs or educational attainment. So what explains the difference? Much evidence suggests that it is partially due to the reverberation of social interactions: as criminals act in a given place and time, they increase the likelihood that others nearby will commit crimes, so that even more crimes occur than would otherwise be expected.[10] And the groups over which these effects extend can number in the hundreds.

A detailed study of these effects by economist Ed Glaeser and his colleagues also shows that certain crimes spread more easily than others, just as one would expect if social influences were more important than local socioeconomic conditions. People are much more likely to be influenced to steal a car when someone else does than to commit a burglary or robbery, and influence is even weaker for crimes like rape and arson. The riskier or more serious the crime, the less likely others are to follow suit (though there can be frenzies of murder too, as in the Rwandan genocide). Moreover, as a further illustration of the social nature of crime, nearly two-thirds of all criminals commit crimes in collaboration with someone else.[11]

While we are not aware of any experimental efforts to foster crime by exploiting social contagion, there have been experiments to study less extreme unethical behaviors. At Carnegie Mellon University, a group of students were asked to take a difficult math test. In the middle of the room, researchers placed a confederate who at some point visibly cheated on the test. When students witnessed the cheater's behavior, they too began to cheat.[12] Especially relevant, though, was the discovery that cheating only increased if the cheater was a person to whom the other students felt connected. If the cheater wore a plain T-shirt, students were more likely to cheat than if he wore a T-shirt from the University of Pittsburgh (Carnegie Mellon's local rival institution).

The Spread of Goodness

In spite of these potential negative effects, we are all connected for a reason. The purpose of social networks is to transmit positive and desirable outcomes, whether joy, warnings about predators, or introductions to romantic partners. To some extent, the transmission of bad behaviors and other adverse phenomena (like germs) are merely side effects that we must endure in order to reap the benefits of networks; they are grafted onto an apparatus that was built, evolutionarily speaking, for another, more beneficial purpose.

To be clear, we are not suggesting a linear progression across history or evolutionary time from anarchy to state to utopia. But we do believe that there is a utopian impulse to form networks that has always been with us. We gain more than we lose by living within social networks, and this drives us to embed ourselves in the lives of others. The natural advantages of a connected life explain why social networks have persisted and why we have come to form a human superorganism.

Crucial traits and behaviors that lie at the root of—and that nourish—social connections have a genetic basis. Altruism, for example, is a key predicate for the formation and operation of social networks. If people never behaved altruistically, never reciprocated kind behavior, or, worse, were always violent, then social ties would dissolve, and the network around us would disintegrate. Some degree of altruism and reciprocity, and indeed some degree of positive emotions such as love and happiness, are therefore crucial for the emergence and endurance of social networks. Moreover, once networks are established, altruistic acts—from random acts of kindness to cascades of organ donation—can spread through them.

Charity is just one example of the goodness that flows through networks. About 89 percent of American households give to charity each year (the average annual contribution was $1,620 in 2001), and fund-raising efforts often seem designed to capitalize on processes of social influence and notions of community embeddedness. Appeals are commonly organized so that people you feel connected to rather than strangers call you to ask for money, such as graduates of your college or relatives of your friend with cancer (of course, it is cheaper to use such volunteers too). Bikeathons and walkathons are organized to engender a sense of community among those participating and to encourage direct contact between participants and the friends and neighbors who sponsor them. And organizations from hospitals to Boy Scout troops to small towns employ a kind of thermometer that is publicly displayed and that tracks charitable giving to their cause, implicitly saying, Look, all these other people gave money; now how about you? Indeed, surveys of people who have given money to diverse causes find that roughly 80 percent did so because they were asked to by someone they knew well.[13]

In one demonstration of the spread of prosocial norms, economist Katie Carman studied charitable giving (via payroll deductions to the United Way) in 2000 and 2001 among the seventy-five thousand employees in a large American bank operating in twenty states. She found that employees gave more when they worked next to generous colleagues. Carman acquired detailed information about the employees' connections at work and their specific locations in bank offices. In a clever exploitation of the most mundane piece of information imaginable—the mail codes used to deliver letters and parcels to areas within bank buildings—she was able to identify groups of people ranging in size from one to 537, with a median size of just nineteen people. She studied what happened to employee giving if they were obliged to move from one location in the bank to another. She found that when people were transferred from a location where others did not give much money to a location where they did, every $1.00 increase in the average giving of their nearby coworkers resulted in a $0.53 increase in their own contribution.[14] There are, of course, several possible mechanisms for this influence: one person could provide information about how to give, could pressure the other to give, or could simply act as a role model for giving.

While Carman's work suggested the person-to-person spread of altruistic norms, our own experiments illustrate the surprising pay-it-forward properties of altruism. We know that if Jay is generous to Harla, Harla will be generous to Jay, but if Jay is generous to Harla, will Harla be generous to Lucas? We devised an experiment to evaluate the idea that altruism could spread from person to person to person. We recruited 120 students for a set of cooperation games that lasted five rounds. In each round, students were placed in groups of four, and we adjusted the composition of the groups so that no two students were ever in the same group twice. Students were each given some funds, and they could decide how much money to give to the group at a personal cost, and then at the end of each round we let them know what the others had done.

When we analyzed their behavior, we found that altruism tends to spread and that the benefits tend to be magnified. When one person gives an extra dollar in the first round, the people in her group each tend to give about twenty cents more in the second round, even though they have been placed in completely new and different groups! When a person has been treated well by someone, she goes on to treat others well in the future. And, even more strikingly, all the people in these new second-round groups are also affected in the third round, each giving about five cents more for every extra dollar that the generous person in the first round spent. Since each group contains three new people at every stage, this means that giving an extra dollar initially caused a total increase in giving by others of sixty cents in the second round and forty-five cents in the third. In other words, the social network acted like a matching grant, prompting an extra $1.05 in total future giving by others for each dollar a person initially chose to give.

Whether people behave altruistically is also determined by the structure of the social network. One ingenious experiment documented a "law of giving" at an all-girls school in Pasadena, California.[15] The investigators asked seventy-six fifth- and sixth-grade girls to identify up to five friends; this allowed investigators to draw the girls' social networks and ascertain which girls were each girl's friends, friends of friends, friends of friends of friends, and so on. They had the girls play the dictator game discussed in chapter 7, and each girl was asked how much she would share from a $6 sum with each of ten other girls who were listed by name. The girls were most generous with their friends, and the amount given declined as social distance increased. On average, the girls offered their friends 52 percent of the $6, friends of friends 36 percent, and friends of friends of friends 16 percent. The best predictor of how much each girl gave was not any measured characteristic of either the givers or the recipients — such as whether either girl was tall or short, had many or few siblings, or wore glasses or braces. Instead, it was the degree of separation between the giver and the receiver.

This is one way that popularity is beneficial. If you are in the center of a social network, you are more likely to be one, two, or three degrees removed from many other people than if you were at the periphery of the network. Consequently, you can earn a centrality premium if good things (like money or respect) are flowing through the network. More people are willing to act altruistically toward you than toward those at the margins. When all the rounds of the game among the schoolgirls were completed, the most popular girls earned four times as much as the least popular. The ability of social networks to magnify whatever they are seeded with favors some people over others.

A pair of experiments with college undergraduates added a few wrinkles to these results.[16] One elicited information about the close friends of 569 undergraduates residing in two large college dormitories in 2003. The other involved 2,360 students using Facebook in 2004. The students were less and less generous to people farther away in the network and were no more generous to people beyond three degrees of separation than they were to total strangers. The college students were also more likely to act altruistically, and to give generously, to social contacts with whom they shared many friends in common. Katrina is more likely to act altruistically toward Dave if they share Ronan and Maddox as friends than if they just share Ronan.

Moreover, the motivation to give to friends that subjects did not expect to interact with again was twice as strong as the motivation to give to strangers that subjects expected to have further interactions with. Put another way, we would rather give a gift to a friend who will never repay us than to give a gift to a stranger who will. The reason is that we give to sustain the network, and it is the network itself that we value. Our social ties repay us for our gifts. Generosity binds the network together, but the network also functions to foster and determine generosity.

This study of college students confirmed a final, crucial point: in real-life interactions, as predicted by the theoretical models discussed in chapter 7, cooperators tend to hang out with other cooperators, and there is homophily in the inclination to be altruistic. Altruistic and selfish undergraduates each had the same number of friends, on average. But altruistic people were embedded in networks of other altruistic people.

Haves and Have-Nots: Social Network Inequality

Today it is common to focus on inequalities in our society that appear to arise from race, income, gender, or geography. We pay attention to the fact that people with better education generally have better health or more economic opportunities, that whites may enjoy advantages that minorities do not, and that where people live affects their life prospects. Politicians, activists, philanthropists, and critics are driven by the recognition that we do not all appear to have equal access to societal goods and that the pattern of access is often manifestly unjust. In short, we live in a hierarchical society, and our sociodemographic characteristics stratify and divide us.

But there is an alternative way of understanding stratification and hierarchy that is based on how people are positioned with respect to their connections. *Positional inequality* occurs not because of who we are but because of who we are connected to. These connections affect where we come to be located in social networks, and they often matter more than our race, class, gender, or education. Some of us have more connections, and some fewer. Some of us are more centrally located, and some of us find ourselves at the periphery. Some of us have densely interconnected social ties and all our friends know one another, and some of us inhabit worlds where none of our friends get along. And these differences are not always of our own doing because our network position also depends on the choices that others around us make.

Not everyone can tap the public goods that are fostered and created by social networks. Your chance of dying after a heart attack may depend more on whether you have friends than on whether you are black or white. Your chance of finding a new job may have as much to do with the friends of your friends as with your skill set. And your chance of being treated kindly or altruistically depends on how well connected others around you are.

Social scientists and policy makers have neglected this kind of inequality, in part because it is so difficult to measure. We cannot understand positional inequality by just studying individuals or even groups. We cannot ask a person where he is located in the social network as easily as we can ask him how much money he earns. Instead, we must observe the social network as a whole before we can understand a person's place in it. This is not a trivial problem. Thankfully, as discussed in chapter 8, the advent of digital communications (e-mail, mobile phones, social-network websites) is making it easier to see networks on a large scale without necessarily surveying individuals at great expense. Correlating people's network centrality with their mortality risk, their transitivity with their prospects of repaying a loan, or their network position with their propensity to commit crimes or quit smoking offers new avenues for policy intervention.

But in an increasingly interconnected world, people with many ties may become even better connected while those with few ties may get left farther and farther behind. As a result, rewards may flow even more toward those with particular locations in social networks. This is the real digital divide. Network inequality creates and reinforces inequality of opportunity. In fact, the tendency of people with many connections to be connected to other people with many connections distinguishes social networks from neural, metabolic, mechanical, or other nonhuman networks. And the reverse holds true as well: those who are poorly connected usually have friends and family who are themselves disconnected from the larger network.

To address social disparities, then, we must recognize that our connections matter much more than the color of our skin or the size of our wallets. To address differences in education, health, or income, we must also address the personal connections of the people we are trying to help. To reduce crime, we need to optimize the kinds of connections potential criminals have—a challenging proposition since we sometimes need to detain criminals. To make smoking-cessation and weight-loss interventions more effective, we need to involve family, friends, and even friends of friends. To reduce poverty, we should focus not merely on monetary transfers or even technical training; we should help the poor form new relationships with other members of society. When we target the periphery of a network to help people reconnect, we help the whole fabric of society, not just any disadvantaged individuals at the fringe.

One for All and All for One

The old ways of understanding human behavior are not up to the task. One classic method used to understand collective human behavior examines the choices and actions of individuals. For instance, we can see markets, elections, and riots as mere by-products of individuals' decisions to buy and sell, cast a ballot, or express anger. The classic example of this approach, which is known as *methodological individualism,* is provided by Adam Smith's conception of markets as the simple sum of individuals' willingness to supply or demand a good.

Another classic method used to understand collective human behavior dispenses with individuals and focuses exclusively on groups delineated by, say, class or race, each with collective identities that cause people in these groups to act in concert. Some scholars in this tradition, such as Karl Marx, believe that groups have their own "consciousness," imbuing them with an indivisible personality that cannot be deduced or understood from the actions of its members. Others have focused on the primacy of group culture. For example, sociologist Émile Durkheim argued that the relatively constant rates of suicide among members of different religious groups across time could not be explained by the actions of any individuals since the groups had an enduring reality that long outlasted the lives of their members. How was it, he wondered, that people came and went, but the suicide rate in French Protestants stayed the same? Known as *methodological holism*, this approach sees social phenomena as having a totality that is distinct from individuals and that cannot be understood by merely studying individuals.

Individualism and holism shed light on the human condition, but they miss something essential. In contrast to these two traditions, the science of social networks offers an entirely new way of understanding human society because it is about individuals *and* groups and, indeed, about how the former become the latter. Interconnections between people give rise to phenomena that are not present in individuals or reducible to their solitary desires and actions. Indeed, culture itself is one such phenomenon. When we lose our connections, we lose everything.

The study of social networks is, in fact, part of a much broader assembly project in modern science. For the past four centuries, swept up by a reductionistic fervor and by considerable success, scientists have been purposefully examining ever-smaller bits of nature in order to understand the whole. We have disassembled life into organs, then cells, then molecules, then genes. We have disassembled matter into atoms, then nuclei, then subatomic particles. We have invented everything from microscopes to supercolliders. But across many disciplines, scientists are now trying to put the parts back together—whether macromolecules into cells, neurons into brains, species into ecosystems, nutrients into foods, or people into networks. Scientists are also increasingly seeing events like earthquakes, forest fires, species extinctions, climate change, heartbeats, revolutions, and market crashes as bursts of activity in a larger system, intelligible only when studied in the context of many examples of the same phenomenon. They are turning their attention to how and why the parts fit together and to the rules that govern interconnection and coherence. Understanding the structure and function of social networks and understanding the phenomenon of emergence (that is, the origin of collective properties of the whole not found in the parts) are thus elements of this larger scientific movement.

Better understanding of social networks is essential for facing new threats in our world. Turmoil in financial markets reminds us that economic activity is becoming increasingly globalized and increasingly interconnected. Emerging public health problems like drug-resistant pathogens and epidemics of risky behaviors are exacerbated by person-to-person spread. Political campaigns are taking greater advantage of new networking technologies, and more and more of our political life is taking place in a hyperconnected world; but these same technologies are used by a few extremists who would like to undo the very world that allows us to connect so well.

All of these challenges require us to recognize that although human beings are individually powerful, we must act together to achieve what we could not accomplish on our own. We have done it before—taming huge rivers, building great cities, creating libraries of knowledge, and sending ourselves into space. We have done it without even knowing all the other people we worked with to make it happen. The miracle of social networks in the modern world is that they unite us with other human beings and give us the capacity to cooperate on a scale so much larger than the one experienced in our ancient past.

But on a more human level, social networks affect every aspect of our lives. Events occurring in distant others can determine the shape of our lives, what we think, what we desire, whether we fall ill or die. In a social chain reaction, we respond to faraway events, often without being consciously aware of it.

Embedded in social networks and influenced by others to whom we are tied, we necessarily lose some of our individuality. Focusing on network connections lessens the importance of individuals in understanding the behaviors of groups. In addition, networks influence many behaviors and outcomes that have moral overtones. If showing kindness and using drugs are contagious, does this mean that we should reshape our own social networks in favor of the benevolent and the abstemious? If we unconsciously copy the good deeds of others to whom we are connected, do we deserve credit for those deeds? And if we adopt the bad habits or evil thoughts of others to whom we are closely or even loosely tied, do we deserve blame? Do they? If social networks place constraints on the information and opinions we have, how free are we to make choices?

Recognition of this loss of self-direction can be shocking. But the surprising power of social networks is not just the effect others have on us. It is also the effect *we* have on others. You do not have to be a superstar to have this power. All you need to do is connect. The ubiquity of human connection means that each of us has a much bigger impact on others than we can see. When we take better care of ourselves, so do many other people. When we practice random acts of kindness, they can spread to dozens or even hundreds of other people. And with each good deed, we help to sustain the very network that sustains us.

The great project of the twenty-first century—understanding how the whole of humanity comes to be greater than the sum of its parts—is just beginning. Like an awakening child, the human superorganism is becoming self-aware, and this will surely help us to achieve our goals. But the greatest gift of this awareness will be the sheer joy of self-discovery and the realization that to truly know ourselves, we must first understand how and why we are all connected.

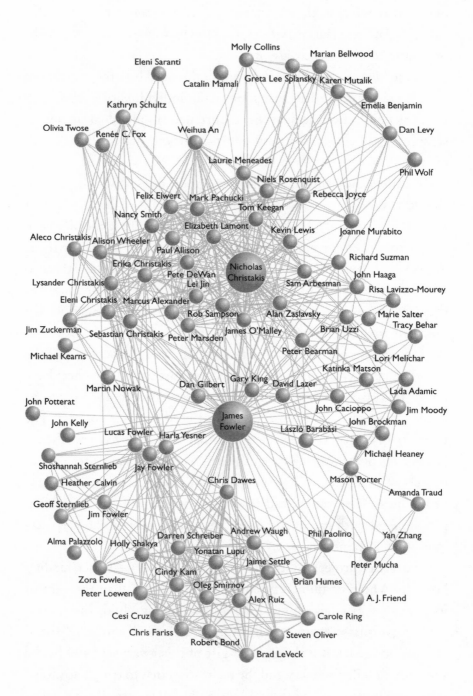

Acknowledgments

These acknowledgments are yet another illustration of the surprising power of social networks. Numerous people to whom we are connected played a critical role in the emergence of this book.

Gary King initiated a long chain of introductions that connected people who were previously several degrees removed. As James's adviser at Harvard, Gary knew James's work on contagion effects in politics. Gary also was friends with Nicholas and knew his work on contagion effects in health, and he spotted a perfect opportunity for multidisciplinary arbitrage. Nicholas's friend and colleague Dan Gilbert ultimately provided extremely creative and valuable comments on early drafts of the book, but the first (and crucial) thing he did was to introduce us to Katinka Matson and John Brockman, who later became our agents. Katinka and John gave us an opportunity to communicate our ideas in a way we had never thought possible. And they, in turn, introduced us to Tracy Behar, our terrific editor at Little, Brown, who played the roles of thoughtful critic and cheerleader.

For brilliant contributions to the literature on networks, and for being first-rate interlocutors, we thank László Barabási, Peter Bearman, David Lazer, and Brian Uzzi. We describe their work in this book, and each of them gave us important advice about the issues we address here.

A variety of friends, colleagues, and relatives read early drafts of parts of this book and gave us valuable comments, including Marcus Alexander, Sam Arbesman, Heather Calvin, Felix Elwert, Michael Heaney, Tom Keegan, Mark Pachucki, Kathryn Schultz, Holly Shakya, Geoff Sternlieb, Shoshannah Sternlieb, and Jim Zuckerman. Nicholas's dissertation adviser, Renée C. Fox, continues to be a force in his life and offered her inimitable and profound perspective on the shape of the argument.

We also owe a debt of intellectual gratitude to Paul Allison, Weihua An, John Cacioppo, Chris Dawes, Pete DeWan, A. J. Friend, Lei Jin, Cindy Kam, Elizabeth Lamont, Peter Marsden, Peter Mucha, Martin Nowak, James O'Malley, Mason Porter, Niels Rosenquist, Rob Sampson, Darren Schreiber, Amanda Traud, Alan Zaslavksy, and Yan Zhang for collaborating with us on various research projects that we describe in this book or for informing our thinking. We thank Lada Adamic, Michael Kearns, John Kelly, Catalin Mamali, Jim Moody, and John Potterat for beautiful versions of their published and unpublished images that we adapted and presented here. And for tireless research assistance and emerging collaborations, we thank Robert Bond, Cesi Cruz, Chris Fariss, Brad LeVeck, Peter Loewen, Yonatan Lupu, Steven Oliver, Alex Ruiz, Jaime Settle, Olivia Twose, Andrew Waugh, and especially Alison Wheeler. For superb copyediting, we thank Marie Salter.

James particularly thanks Chris Dawes for teaching him to be a better scientist and taking him to the best Major League Baseball game of his life. He is also especially indebted to Oleg Smirnov, his first collaborator, for getting him to think about altruistic punishment and evolutionary psychology, and to the Santa Fe Institute (where they met) for encouraging people to think outside their disciplines.

None of our work would be possible without generous financial and logistical support. Richard Suzman and John Haaga at the National Institute on Aging, and Brian Humes and Phil Paolino at the National Science Foundation, took a chance on social network and behavior genetics research. Risa Lavizzo-Mourey and Lori Melichar at the Robert Wood Johnson Foundation, through its Pioneer program, also provided early support for our data development and methodological efforts. Emelia Benjamin, Dan Levy, Joanne Murabito, Karen Mutalik, Greta Lee Splansky, and Phil Wolf supported our work with the Framingham Heart Study (a national treasure), and Marian Bellwood introduced us to the paper "green sheets" that contained all that valuable, previously unused social network information. Laurie Meneades and Rebecca Joyce (and, earlier, Molly Collins) offered us superb data support with the Framingham social network, and Kevin Lewis did a valuable job spearheading the Facebook data-collection effort. And we thank Alma Palazzolo, Carole Ring, and Nancy Smith for their energetic administrative support; we also thank Nancy for her patience.

Finally, like the kings who arrive onstage at the end of Shakespearean plays, the most important people in our lives appear here. Nicholas thanks Aleco Christakis for valuable comments and for his key connection to Eleni Saranti. Nicholas also thanks Erika Christakis for critically reading the book and improving it substantially. His gratitude extends well beyond this book: everything good in his life is connected to her. Eleni, Lysander, and Sebastian had all sorts of valuable things to say about texting and Facebook and World of Warcraft and much else that their father cannot immediately grasp.

James thanks Jim Fowler for valuable comments, Zora Fowler for unbridled enthusiasm, and both of them for tremendous social support and the perfect representation of what it takes to create an enduring connection. He also thanks Harla Yesner for reading several chapters, enduring obsessive dinner conversations about the content of the book, and for always being the perfect partner. Jay Fowler and Lucas Fowler also gave him some extra time on the weekends when he should have been playing Wii Super Smash Bros. Brawl.

Thank you all.

Notes

CHAPTER 1: IN THE THICK OF IT

1. R. V. Gould, "Revenge as Sanction and Solidarity Display: An Analysis of Vendettas in Nineteenth-Century Corsica," *American Sociological Review* 65 (2000): 682–704.
2. B. A. Jacobs, "A Typology of Street Criminal Retaliation," *Journal of Research in Crime and Delinquency* 41 (2004): 295–323.
3. A. V. Papachristos, "Murder by Structure: Dominance Relations and the Social Structure of Gang Homicide in Chicago," December 2007, http://ssrn.com/abstract=855304.
4. M. Planty, *Third Party Involvement in Violent Crime, 1993–1999* (Washington, DC: Bureau of Justice Statistics Special Report, 2002).
5. www.livingdonorsonline.com (accessed October 3, 2008).
6. "Walk Strikes Close to Home for Organ Donor's Family," *The Mississauga News*, April 28, 2008.
7. M. A. Rees and others, "A Nonsimultaneous, Extended, Altruistic-Donor Chain," *New England Journal of Medicine* 360 (2009): 1096–1101.
8. "'Moon Tracking' Station Readied at Canoga Park," *Los Angeles Times*, November 25, 1957.
9. P. Marsden, "Core Discussion Networks of Americans," *American Sociological Review* 52 (1987): 122–31; see also: M. McPherson and others, "Social Isolation in America: Changes in Core Discussion Networks Over Two Decades," *American Sociological Review* 71 (2006): 353–75.
10. I. de Sola Pool and M. Kochen, "Contacts and Influence," *Social Networks* 1 (1978/1979): 5–51.
11. P. Kristensen and T. Bjerkedal, "Explaining the Relation Between Birth Order and Intelligence," *Science* 316 (2007): 1717.
12. S. Milgram, L. Bickman, and L. Berkowitz, "Note on the Drawing Power of Crowds of Different Size," *Journal of Personality and Social Psychology* 13 (1969): 79–82.
13. I. Farkas and others, "Mexican Waves in an Excitable Medium," *Nature* 419 (2002): 131–32.
14. I. D. Couzin and others, "Effective Leadership and Decision-Making in Animal Groups on the Move," *Nature* 433 (2005): 513–16.
15. J. Travers and S. Milgram, "An Experimental Study in the Small World Problem," *Sociometry* 35, no. 4 (1969): 425–43.
16. P. S. Dodds and others, "An Experimental Study of Search in Global Social Networks," *Science* 301 (2003): 827–29.

CHAPTER 2: WHEN YOU SMILE, THE WORLD SMILES WITH YOU

1. A. M. Rankin and P. J. Philip, "An Epidemic of Laughing in the Bukoba District of Tanganyika," *Central African Journal of Medicine* 9 (1963): 167–70.
2. A. Hatfield and others, "Emotional Contagion," *Current Directions in Psychological Science* 2 (1993): 96–99.
3. C. N. Scollon and others, "Experience Sampling: Promise and Pitfalls, Strengths and Weaknesses," *Journal of Happiness Studies* 4 (2003): 5–34; J. P. Laurenceau and N. Bolger, "Using Diary Methods to Study Marital and Family Processes," *Journal of Family Psychology* 19 (2005): 86–97; R. Larson and M. H. Richards, *Divergent Realities: The Emotional Lives of Mothers, Fathers, and Adolescents* (New York: Basic Books, 1994).
4. M. J. Howes and others, "Induction of Depressive Affect After Prolonged Exposure to a Mildly Depressed Individual," *Journal of Personality and Social Psychology* 49 (1985): 1110–13.
5. S. D. Pugh, "Service with a Smile: Emotional Contagion in the Service Encounter," *Academy of Management Journal* 44 (2001): 1018–27; W. C. Tsai and Y. M. Huang, "Mechanisms Linking Employee Affective Delivery and Customer Behavioral Intentions," *Journal of Applied Psychology* 87 (2002):1001–8.
6. J. B. Silk, "Social Components of Fitness in Primate Groups." *Science* 317 (2007): 1347–51.
7. J. M. Susskind and others, "Expressing Fear Enhances Sensory Acquisition," *Nature Neuroscience* 11 (2008): 843–50.
8. Cited in Hatfield, "Emotional Contagion," 97.
9. M. Iacoboni, *Mirroring People: The New Science of How We Connect with Others* (New York: Farrar, Straus, and Giroux, 2008).
10. J. E. Warren and others, "Positive Emotions Preferentially Engage an Auditory-Motor 'Mirror' System," *Journal of Neuroscience* 26 (2006): 13067–75.
11. J. F. C. Hecker, *The Epidemics of the Middle Ages,* trans. by B. G. Babington (London: Sydenham Society, 1844), 87–88.
12. T. F. Jones and others, "Mass Psychogenic Illness Attributed to Toxic Exposure at a High School," *New England Journal of Medicine* 342 (2000): 96–100.
13. "Mass Hysteria Can Make Many Sick—Tennessee Case Shows How Anxiety Spreads," *Florida Times-Union,* May 7, 2000.
14. Jones, "Mass Psychogenic Illness," 100.
15. L. P. Boss, "Epidemic Hysteria: A Review of the Published Literature," *Epidemiological Reviews* 19 (1997): 233–43.
16. D. M. Johnson, "The 'Phantom Anesthetist' of Mattoon: A Field Study of Mass Hysteria,' in *Readings in Social Psychology* (New York: Henry Holt and Co., 1952), 210.
17. E. T. Rolls, "The Functions of the Orbitofrontal Cortex," *Brain and Cognition* 55 (2004): 11–29.
18. J. Willander and M. Larsson, "Olfaction and Emotion: The Case of Autobiographical Memory," *Memory and Cognition* 35 (2007): 1659–63.
19. R. S. Herz and others, "Neuroimaging Evidence for the Emotional Potency of Odor-Evoked Memory," *Neuropsychologia* 42 (2004): 371–78; D. H. Zald and J. V. Pardo, "Emotion, Olfaction, and the Human Amygdala: Amygdala Acti-

vation During Aversive Olfactory Stimulation," *Proceedings of the National Academy of Sciences* 94 (1997): 4119–24.

20. Boss, "Epidemic Hysteria," 238.

21. M. Talbot, "Hysteria Hysteria," *New York Times,* June 2, 2002.

22. N. A. Christakis, "This Allergies Hysteria Is Just Nuts," *British Medical Journal* 337 (2008): a2880.

23. M. Csikszentmihalyi and R. Larson, "Validity and Reliability of the Experience-Sampling Method," *Journal of Nervous and Mental Diseases* 175 (1987): 527; M. Csikszentmihalyi and others, "The Ecology of Adolescent Activity and Experience," *Journal of Youth and Adolescence* 6 (1977): 281–94.

24. R. W. Larson and M. H. Richards, "Family Emotions: Do Young Adolescents and Their Parents Experience the Same States?" *Journal of Research on Adolescence* 4 (1994): 567–83.

25. P. Totterdell, "Catching Moods and Hitting Runs: Mood Linkage and Subjective Performance in Professional Sports Teams," *Journal of Applied Psychology* 85 (2000): 848–59.

26. J. H. Fowler and N. A. Christakis, "Dynamic Spread of Happiness in a Large Social Network: Longitudinal Analysis Over 20 Years in the Framingham Heart Study," *British Medical Journal* 337 (2008): a2338.

27. R. A. Easterlin, "Explaining Happiness," *Proceedings of the National Academy of Sciences* 100 (2003): 11176–83.

28. J. Knight and R. Gunatilaka, "Is Happiness Infectious?" (unpublished paper, Oxford University, 2009).

29. E. Diener, R. E. Lucas, and C. N. Scollon, "Beyond the Hedonic Treadmill: Revising the Adaptation Theory of Well-Being," *American Psychologist* 61 (2006): 305–14.

30. D. Gilbert, *Stumbling on Happiness* (New York: Vintage, 2005).

31. S. Lyubormirsky and others, "Pursuing Happiness: The Architecture of Sustainable Change," *Review of General Psychology* 9 (2005): 111–31.

32. J. M. Ernst and J. T. Cacioppo, "Lonely Hearts: Psychological Perspectives on Loneliness," *Applied and Preventive Psychology* 8, no. 1 (1999): 1–22.

33. J. T. Cacioppo, J. H. Fowler, and N. A. Christakis, "Alone in the Crowd: The Structure and Spread of Loneliness in a Large Social Network," *Journal of Personality and Social Psychology* (forthcoming).

34. H. Fisher, *Why We Love: The Nature and Chemistry of Romantic Love* (New York: Henry Holt and Co., 2004).

CHAPTER 3: LOVE THE ONE YOU'RE WITH

1. http://www.city-data.com/forum/relationships/331411-how-i-met-my-spouse-3.html (accessed March 3, 2009). Reprinted by permission of the author.

2. E. O. Laumann and others, *The Social Organization of Sexuality: Sexual Practices in the United States* (Chicago: University of Chicago Press, 1994).

3. M. Bozon and F. Héran, "Finding a Spouse: A Survey of How French Couples Meet," *Population* 44, no. 1 (1989): 91–121.

4. Laumann, *Social Organization of Sexuality.*

5. Bozon, "Finding a Spouse."

6. M. Madden and A. Lenhart, "Online Dating," Pew Internet and American Life Project, http://www.pewinternet.org/pdfs/PIP_Online_Dating.pdf, ii (accessed February 28, 2009).

7. Ibid.

8. Ibid., iii.

9. X. Xiaohe and M. K. Whyte, "Love Matches and Arranged Marriage: A Chinese Replication," *Journal of Marriage and Family* 52 (1990): 709–22; see also N. P. Medora, "Mate Selection in Contemporary India: Love Marriages Versus Arranged Marriages," in *Mate Selection Across Cultures*, ed. H. R. Hamon and B. B. Ingoldsby (Thousand Oaks, CA: Sage Publications, 2003): 209–30.

10. Laumann, *Social Organization of Sexuality*, 255.

11. J. K. Galbraith, *The Affluent Society* (Boston, MA: Houghton Mifflin, 1958).

12. A. Tverksy and D. Griffin, *Strategy and Choice* (Cambridge, MA: Harvard University Press, 1991).

13. S. J. Solnick and D. Hemenway, "Is More Always Better? A Survey on Positional Concerns," *Journal of Economic Behavior and Organization* 37 (1998): 373–83.

14. L. Jin and others, "Reduction in Long-Term Survival in Men Given High Operational Sex Ratio at Sexual Maturity," *Demography* (forthcoming).

15. R. K. Merton, *Social Theory and Social Structure* (New York: Free Press of Glencoe, 1957); R. K. Merton and A. S. Kitt, "Contributions to the Theory of Reference Group Behavior," in *Continuities in Social Research: Studies in the Scope and Method of "The American Soldier,"* ed. R. K. Merton and P. F. Lazarsfeld (Glencoe, IL: Free Press, 1950), 40–105; A. Bandura, *Social Learning Theory* (New York: General Learning Press, 1971); L. Festinger, "A Theory of Social Comparison Processes," *Human Relations* 7 (1954): 117–40.

16. B. C. Jones and others, "Social Transmission of Face Preferences Among Humans," *Proceedings of the Royal Society B* 274 (2007): 899–903.

17. K. Eva and T. Wood, "Are All The Taken Men Good? An Indirect Examination of Mate-Choice Copying in Humans," *Canadian Medical Association Journal* 175 (2006): 1573–74.

18. D. Waynforth, "Mate Choice Copying in Humans," *Human Nature* 18 (2007): 264–71.

19. P. Bressan and D. Stranieri, "The Best Men Are (Not Always) Already Taken: Female Preference for Single Versus Attached Males Depends on Conception Risk," *Pyschological Science* 19 (2008): 145–51.

20. D. T. Gilbert and others, "The Surprising Power of Neighborly Advice," *Science* 323 (2009): 1617–19.

21. S. E. Hill and D. M. Buss, "The Mere Presence of Opposite-Sex Others on Judgments of Sexual and Romantic Desirability: Opposite Effects for Men and Women," *Personality and Social Psychology Bulletin* 34 (2008): 635–47.

22. M. D. Regnerus and L. B. Luchies, "The Parent-Child Relationship and Opportunities for Adolescents' First Sex," *Journal of Family Issues* 27 (2006): 159–83.

23. S. E. Cavanagh, "The Sexual Debut of Girls in Early Adolescence: The Intersection of Race, Pubertal Timing, and Friendship Group Characteristics," *Journal of Research on Adolescence* 14 (2004): 285–312.

24. A. Adamczyk and J. Felson, "Friends' Religiosity and First Sex," *Social Science Research* 35 (2006): 924–47.

25. P. S. Bearman and H. Brückner, "Promising the Future: Abstinence Pledges and the Transition to First Intercourse," *American Journal of Sociology* 106 (2001): 859–912.

26. W. Manning, M. A. Longmore, and P. C. Giordano, "Adolescents' Involvement in Non-Romantic Sexual Activity," *Social Science Research* 34 (2005): 384–407.

27. M. J. Prinstein, C. S. Meade, and G. L. Cohen, "Adolescent Oral Sex, Peer Popularity, and Perceptions of Best Friend's Sexual Behavior," *Journal of Pediatric Psychology* 28 (2003): 243–49.

28. Laumann, *Social Organization of Sexuality*.

29. I. Kuziemko, "Is Having Babies Contagious? Estimating Fertility Peer Effects BetweenSiblings,"http://www.princeton.edu/~Ekuziemko/fertility_11_29_06.pdf (accessed March 1, 2009).

30. D. E. Bloom and others, "Social Interactions and Fertility in Developing Countries," PGDA *Working Paper* 34 (2008).

31. W. Farr, "Influence of Marriage on the Mortality of the French People," in *Transactions of the National Association for the Promotion of Social Science*, ed. G. W. Hastings (London: John W. Park & Son, 1858), 504–13.

32. D. Lubach, quoted in F. Van Poppel and I. Joung, "Long Term Trends in Marital Status Differences in the Netherlands 1850–1970," *Journal of Biosocial Science* 33 (2001): 279–303.

33. B. Turksma, quoted in Van Poppel and I. Joung, "Long Term Trends in Marital Status Differences in the Netherlands 1850–1970," *Journal of Biosocial Science* 33 (2001): 279–303.

34. M. Young, B. Benjamin, and C. Wallis, "The Mortality of Widowers," *Lancet* 2, no. 7305 (1963): 454–56.

35. F. Elwert and N. A. Christakis, "Variation in the Effect of Widowhood on Mortality by the Causes of Death of Both Spouses," *American Journal of Public Health* 98 (2008): 2092–98.

36. L. J. Waite, "Does Marriage Matter?" *Demography* 32 (1995): 483–508.

37. L. A. Lillard and L. J. Waite, " 'Til Death Do Us Part—Marital Disruption and Mortality," *American Journal of Sociology* 100 (1995): 1131–56; L. A. Lillard and C. A. W. Panis, "Marital Status and Mortality: The Role of Health," *Demography* 33 (1996): 313–27.

38. See, for example: K. Allen, J. Blascovich, W. B. Mendes, "Cardiovascular Reactivity and the Presence of Pets, Friends, and Spouses: The Truth about Cats and Dogs," *Psychosomatic Medicine* 64 (2002): 727–39; J. K. Kiecolt-Glaser and others, "Marital Quality, Marital Disruption, and Immune Function," *Psychosomatic Medicine* 49 (1987): 13–34.

39. T. J. Iwashyna and N. A. Christakis, "Marriage, Widowhood, and Health Care Use," *Social Science and Medicine* 57 (2003): 2137–47; L. Jin and N. A. Christakis, "Investigating the Mechanism of Marital Mortality Reduction: The Transition to Widowhood and Quality of Health Care," *Demography* (forthcoming).

40. D. Umberson, "Family Status and Health Behaviors: Social Control as a Dimension of Social Integration," *Journal of Health and Social Behavior* 28 (1987): 306–19; D. Umberson, "Gender, Marital Status and the Social Control of Health Behavior," *Social Science and Medicine* 34 (1992): 907–17.

41. F. Elwert and N. A. Christakis, "Widowhood and Race," *American Sociological Review* 71 (2006): 16–41.
42. See, for example, ibid.
43. Y. Hu and N. Goldman, "Mortality Differentials by Marital Status: An International Comparison," *Demography* 27 (1990): 233–50.
44. Elwert, "Widowhood and Race."
45. D. Umberson and others, "You Make Me Sick: Marital Quality and Health Over the Life Course," *Journal of Health and Social Behavior* 47 (2006): 1–16; see also D. Carr, "Gender, Pre-loss Marital Dependence, and Older Adults' Adjustment to Widowhood," *Journal of Marriage and Family* 66 (2004): 220–35.
46. G. Clark, *Too Brief a Treat—Letters of Truman Capote* (New York: Random House, 2004).

CHAPTER 4: THIS HURTS ME AS MUCH AS IT HURTS YOU

1. http://www.rockdalecounty.org/main.cfm?id=2130 (accessed March 1, 2009).
2. M. A. J. McKenna, "Teen Sex Tales Turn National Focus to Rockdale," *Atlanta Journal-Constitution*, October 19, 1999; see also R. B. Rothenberg and others, "Using Social Network and Ethnographic Tools to Evaluate Syphilis Transmission," *Sexually Transmitted Diseases* 25 (1998): 154–60.
3. McKenna, "Teen Sex Tales."
4. C. Russell, "Venereal Disease Rampant Among America's Teenagers; Health Officials Call for Prevention and Study," *Washington Post*, November 26, 1996.
5. McKenna, "Teen Sex Tales."
6. Rothenberg, "Using Social Network and Ethnographic Tools."
7. P. S. Bearman, J. Moody, and K. Stovel, "Chains of Affection," *American Journal of Sociology* 110 (2004): 44–91.
8. J. J. Potterat and others, "Sexual Network Structure as an Indicator of Epidemic Phase," *Sexually Transmitted Infections* 78 (2002): 152–58.
9. E. O. Laumann and Y. Youm, "Racial/Ethnic Group Differences in the Prevalence of Sexually Transmitted Diseases in the United States: A Network Explanation," *Sexually Transmitted Diseases* 26 (1999): 250–61.
10. F. Liljeros and others, "The Web of Human Sexual Contacts," *Nature* 411 (2001): 908–9.
11. S. Helleringer and H. P. Kohler, "Sexual Network Structure and the Spread of HIV in Africa: Evidence from Likoma Island, Malawi," *AIDS* 21 (2007): 2323–32.
12. H Epstein, *The Invisible Cure: Africa, the West, and the Fight Against AIDS* (New York: Farrar, Straus, and Giroux, 2007).
13. B. Wansink, *Mindless Eating: Why We Eat More Than We Think* (New York: Bantam, 2006); V. I. Clendenen, C. P. Herman, and J. Polivy, "Social Facilitation of Eating among Friends and Strangers," *Appetite* 23 (1994): 1–13.
14. N. A. Christakis and J. H. Fowler, "The Spread of Obesity in a Large Social Network Over 32 Years," *New England Journal of Medicine* 357 (2007): 370–79.
15. J. H. Fowler and N. A. Christakis, "Estimating Peer Effects on Health in Social Networks," *Journal of Health Economics* 27(2008): 1386–91.
16. P. R. Provine, "Contagious Yawning and Laughing: Everyday Imitation and Mirror-Like Behavior," *Behavioral and Brain Sciences* 28 (2005): 142.
17. D. G. Blanchflower, A. J. Oswald, and B. Landeghem, "Imitative Obesity and Relative Utility," NBER Working Paper W14377 (2008).

18. E. Goodman, "Obesity 'Contagion,'" *Boston Globe,* August 3, 2007.

19. N. A. Christakis and J. H. Fowler, "The Collective Dynamics of Smoking in a Large Social Network," *New England Journal of Medicine* 358 (2008): 2249–58.

20. A. M. Brandt, *The Cigarette Century* (New York: Basic Books, 2007).

21. P. Ormerod and G. Wiltshire, "'Binge' Drinking in the UK: A Social Network Phenomenon," http://arxiv.org/abs/0806.3176 (accessed March 1, 2009).

22. N. Rao, M. M. Mobius, and T. Rosenblat, "Social Networks and Vaccination Decisions," Federal Reserve Bank of Boston working paper #07-12 (2007).

23. H. Raspe, A. Hueppe, and H. Neuhauser, "Back Pain: A Communicable Disease?" *International Journal of Epidemiology* 37 (2008): 69–74.

24. B. F. Walker, "The Prevalence of Low Back Pain: A Systematic Review of the Literature from 1966 to 1998," *Journal of Spinal Disorders* 13 (2000): 205–17.

25. A. R. Lucas and others, "50-Year Trends in the Incidence of Anorexia Nervosa in Rochester, Minn.: A Population-Based Study," *American Journal of Psychiatry* 148 (1991): 917–22; American Psychiatric Association Work Group on Eating Disorders, "Practice Guideline for the Treatment of Patients with Eating Disorders," *American Journal of Psychiatry* 157 (2000): 1–39.

26. C. S. Crandall, "Social Contagion of Binge Eating," *Journal of Personality and Social Psychology* 55 (1988): 588–98.

27. M. S. Gould, S. Wallenstein, and M. Kleinman, "Time-Space Clustering of Teenage Suicide," *American Journal of Epidemiology* 131 (1990): 71–78.

28. D. P. Phillips, "The Influence of Suggestion on Suicide: Substantive and Theoretical Implications of the Werther Effect," *American Sociological Review* 39 (1974): 340–54.

29. Centers for Disease Control, "Suicide Contagion and the Reporting of Suicide: Recommendations from a National Workshop," *Morbidity and Mortality Weekly Review* 43, no. RR-6 (1994): 9–18.

30. E. Etzerdsorfer and G. Sonneck, "Preventing Suicide by Influencing Mass-Media Reporting: The Viennese Experience, 1980–1996," *Archives of Suicide Research* 4 (1998): 67–74.

31. M. S. Gould and others, "Suicide Cluster: An Examination of Age-Specific Effects," *American Journal of Public Health* 80 (1990): 211–12.

32. C. Wilkie, S. Macdonald, and K. Hildahl, "Community Case Study: Suicide Cluster in a Small Manitoba Community," *Canadian Journal of Psychiatry* 43 (1998): 823–28. Reprinted with permission.

33. D. A. Brent and others, "An Outbreak of Suicide and Suicidal Behavior in a High School," *Journal of the American Academy of Child and Adolescent Psychiatry* 28 (1989): 918–24.

34. See, for example: UPI, "Japanese Internet Suicide Clubs Targeted by Police," October 7, 2005; BBC News, "Nine Die in Japan Suicide Pacts," October 12, 2004; and S. Rajagopal, "Suicide Pacts and the Internet," *British Medical Journal* 329 (2004): 1298–99.

35. P. S. Bearman and J. Moody, "Suicide and Friendships Among American Adolescents," *American Journal of Public Health* 94 (2004): 89–95.

36. P. Hedstrom, K. Y. Liu, and M. K. Nordvik, "Interaction Domains and Suicides: A Population-Based Panel Study of Suicides in Stockholm, 1991–1999," *Social Forces* 87 (2008): 713–40.

37. M. D. Resnick and others, "Protecting Adolescents from Harm: Findings from

the National Longitudinal Study of Adolescent Health," *Journal of the American Medical Association* 278 (1997): 823–32.

38. Centers for Disease Control, "Suicide Contagion."

39. M. Gould, P. Jamieson, and D. Romer, "Media Contagion and Suicide Among the Young," *American Behavioral Scientist* 46 (2003): 1269–84.

40. R. R. Wing and R. W. Jeffery, "Benefits of Recruiting Participants with Friends and Increasing Social Support for Weight Loss and Maintenance," *Journal of Consulting and Clinical Psychology* 67 (1999): 132–38.

41. A. A. Gorin and others, "Weight Loss Treatment Influences Untreated Spouses and the Home Environment: Evidence of a Ripple Effect," *International Journal of Obesity* 32 (2008): 1678–84; see also A. L. Shattuck, E. White, A. R. Kristal, "How Women's Adopted Low-Fat Diets Affect Their Husbands," *American Journal of Public Health* 82 (1992): 1244–50; R. S. Zimmerman and others, "The Effects of a Worksite Health Promotion Program on the Wives of Firefighters," *Social Science and Medicine* 26 (1988): 537–43.

42. T. W. Valente and P. Pumpuang, "Identifying Opinion Leaders to Promote Behavior Change," *Health Education and Behavior* 34 (2007): 881–96.

43. D. B. Buller and others, "Randomized Trial Testing the Effect of Peer Education in Increasing Fruit and Vegetable Intake," *Journal of the National Cancer Institute* 91 (1999):1491–1500; K. J. Sikkema and others, "Outcomes of a Randomized Community-Level HIV Prevention Intervention for Women Living in 18 Low-Income Housing Developments," *American Journal of Public Health* 90 (2000): 57–63.

44. D. J. Watts and P. S. Dodds, "Influentials, Networks, and Public Opinion Formation," *Journal of Consumer Research* 34 (2007): 441–58.

45. D. Bahr and others, "Exploiting Social Networks to Mitigate the Obesity Epidemic," *Obesity* 17 (2009): 723–28.

46. R. Cohen, S. Havlin, and D. Aen-Avraham, "Efficient Immunization Strategies for Computer Networks and Populations," *Physical Review Letters* 91 (2003): 247901.

47. J. Leskovec and others, "Cost-Effective Outbreak Detection in Networks," in *Proceedings of the 13th* ACM SIGKDD *International Conference on Knowledge Discovery and Data Mining* (New York: Association for Computing Machinery, 2007), 420–29.

CHAPTER 5: THE BUCK STARTS HERE

1. P. Trowbridge and S. Thompson, "Northern Rock Experiences Second Day of Withdrawals," *Bloomberg*, September 15, 2007.

2. Ibid.

3. "Panic Grips Northern Rock Savers for Second Straight Day," *AFP*, September 13, 2007.

4. B. Livesey and J. Menon, "Northern Rock Stock Tumbles Further Amid Run on Bank," *Bloomberg*, September 17, 2007.

5. "The Great Northern Run," *Economist*, September 20, 2007.

6. M. Oliver, "Customers Rush to Withdraw Money," *Guardian*, September 14, 2007.

7. J. Werdigier, "A Rush to Cash out of Northern Rock," *International Herald Tribune*, September 17, 2007.

8. D. Segal, "In Letter, Buffet Accepts Blame and Faults Others," *New York Times*, March 1, 2009.

9. M. Kelly and C. O'Grada, "Market Contagion: Evidence from the Panics of 1854 and 1857," *American Economic Review* 90 (2000): 1110–24.

10. M. Grabell, "Dallas: Venue Closing for 5 Months After Prostitution Arrests," *Dallas Morning News,* February 6, 2007.

11. S. Scott and C. Duncan, *Biology of Plagues: Evidence from Historical Populations* (Cambridge: Cambridge University Press, 2001).

12. "Web Game Provides Breakthrough in Predicting Spread of Epidemics," http://www.scienceblog.com/cms/web_game_provides_breakthrough_in_predicting_spread_of_epidemics_9874 (accessed March 1, 2009).

13. D. Brockmann, L. Hufnagel, and T. Geisel, "The Scaling Laws of Human Travel," *Nature* 439 (2006): 462–65.

14. R. Shiller, *Irrational Exuberance* (Princeton, NJ: Princeton University Press, 2005).

15. F. Galton, "Vox Populi," *Nature* 75 (1907): 450–51.

16. J. H. Fowler, "Elections and Markets: The Effect of Partisan Orientation, Policy Risk, and Mandates on the Economy," *Journal of Politics* 68 (2006): 89–103; J. H. Fowler and O. Smirnov, *Mandates, Parties, and Voters: How Elections Shape the Future* (Philadelphia: Temple University Press, 2007).

17. K. J. Arrow and others, "The Promise of Prediction Markets," *Science* 320 (2008): 877–78.

18. M. J. Salganik, P. S. Dodds, and D. J. Watts, "Experimental Study of Inequality and Unpredictability in an Artificial Cultural Market," *Science* 311 (2006): 854–56.

19. E. M. Rogers, *Diffusion of Innovations,* 5th ed. (New York: Free Press, 2003).

20. J. J. Brown and P. H. Reingen, "Social Ties and Word of Mouth Referral Behavior," *Journal of Consumer Research* 14 (1987): 350–62.

21. A. B. Jaffe and M. Trajtenberg, *Patents, Citations and Innovations: A Window on the Knowledge Economy* (Cambridge, MA: MIT Press, 2002); E. Duguet and M. MacGarvie, "How Well Do Patent Citations Measure Knowledge Spillovers? Evidence from French Innovative Surveys," *Economics of Innovation and New Technology* 14 (2005): 375–93.

22. J. Singh, "Collaborative Networks as Determinants of Knowledge Diffusion Patterns," *Management Science* 51 (2005): 756–70.

23. M. Granovetter, "The Strength of Weak Ties," *American Journal of Sociology* 78 (1973): 1360–80.

24. J. F. Padgett and C. K. Ansell, "Robust Action and the Rise of the Medici, 1400–1434," *American Journal of Sociology* 98 (1993): 1259–1319.

25. A. J. Hillman, "Politicians on the Board of Directors: Do Connections Affect the Bottom Line?" *Journal of Management* 31 (2005): 464–81.

26. P. Mariolis, "Interlocking Directorates and the Control of Corporations," *Social Science Quarterly* 56 (1975): 425–39.

27. V. Burris, "Interlocking Directorates and Political Cohesion among Corporate Elites," *American Journal of Sociology* 111 (2005): 249–83.

28. B. Uzzi, "The Sources and Consequences of Embeddedness for the Economic Performance of Organizations: The Network Effect," *American Sociological Review* 61 (1996): 674–98.

29. B. Uzzi and J. Spiro, "Collaboration and Creativity: The Small World Problem," *American Journal of Sociology* 111 (2005): 447–504.

30. D. J. Watts, S. H. Strogatz, "Collective Dynamics of 'Small-World' Networks," *Nature* 393 (1998): 409–10.
31. M. Kearns, S. Suri, and N. Montfort, "An Experimental Study of the Coloring Problem on Human Subject Networks," *Science* 313 (2006): 824–27.
32. M. Kearns and others, "Behavioral Experiments on Biased Voting in Networks," *Proceedings of the National Academy of Sciences* 106 (2009): 1347–52.
33. M. Yunus, *Banker to the Poor: Micro-Lending and the Battle Against World Poverty* (New York: Public Affairs, 2003), 62.
34. C. Geertz, "The Rotating Credit Association: A 'Middle Rung' in Development," *Economic Development and Cultural Change* 10 (1962): 241–63.
35. T. Besley, S. Coate, and G. Loury, "The Economics of Rotating Savings and Credit Associations," *American Economic Review* 83 (1993): 792–810; S. Ardner, "The Comparative Study of Rotating Credit Associations," *Journal of the Royal Anthropological Institute* 94 (1964): 202–29.

CHAPTER 6: POLITICALLY CONNECTED

1. A. Smith and L. Raine, "The Internet and the 2008 Election," June 15, 2008, http://www.pewinternet.org/~/media//Files/Reports/2008/PIP_2008_election.pdf (accessed April 4, 2009).
2. A. Downs, *An Economic Theory of Democracy* (New York: Harper, 1957).
3. W. Riker and P. Ordeshook, "A Theory of the Calculus of Voting," *American Political Science Review* 62 (1968): 25–42.
4. A. J. Fischer, "The Probability of Being Decisive," *Public Choice* 101 (1999): 267–83; I. J. Good and L. S. Mayer, "Estimating the Efficacy of a Vote," *Behavioral Science* 20 (1975): 25–33; G. Tullock, *Towards a Mathematics of Politics* (Ann Arbor, MI: University of Michigan, 1967).
5. C. B. Mulligan and C. G. Hunter, "The Empirical Frequency of a Pivotal Vote," *Public Choice* 116 (2003): 31–54.
6. A. Gelman, G. King, and J. Boscardin, "Estimating the Probability of Events That Have Never Occurred: When Is Your Vote Decisive?" *Journal of the American Statistical Association* 93 (1998): 1–9.
7. A. Blais and R. Young, "Why Do People Vote? An Experiment in Rationality," *Public Choice* 99 (1999): 1–2, 39–55.
8. T. Carpenter, "Professor Registers to Vote," *Lawrence Journal-World*, November 12, 1996.
9. M. Fiorina, "Information and Rationality in Elections," in *Information and Democratic Processes*, ed. J. Ferejohn and J. Kuklinski (Urbana, IL: University of Illinois Press, 1990): 329–42.
10. A. Campbell, G. Gurin, and W. E. Miller, *The Voter Decides* (Evanston, IL: Row, Peterson and Company, 1954); W. A. Glaser, "The Family and Voting Turnout," *Public Opinion Quarterly* 23 (1959): 563–70; B. C. Straits, "The Social Context of Voter Turnout," *Public Opinion Quarterly* 54 (1990): 64–73; S. Knack, "Civic Norms, Social Sanctions, and Voter Turnout," *Rationality and Society* 4 (1992): 133–56; C. B. Kenny, "Political Participation and Effects from the Social Environment," *American Journal of Political Science* 36 (1992): 259–67; C. B. Kenny, "The Microenvironment of Political Participation," *American Politics Quarterly* 21 (1993): 223–38; P. A. Beck and others, "The

Social Calculus of Voting: Interpersonal, Media, and Organizational Influences on Presidential Choices," *American Political Science Review* 96 (2002): 57–74.

11. P. F. Lazarsfeld, B. Berelson, and H. Gaudet, *The People's Choice* (New York: Columbia University, 1944); B. Berelson, P. F. Lazarsfeld, and W. N. McPhee, *Voting* (Chicago: University of Chicago Press, 1954).

12. R. Huckfeldt and J. Sprague, *Citizens, Parties, and Social Communication* (New York: Cambridge University Press, 1995).

13. R. Huckfeldt, "Political Loyalties and Social Class Ties: The Mechanisms of Contextual Influence," *American Journal of Political Science* 28 (1984): 414.

14. R. Huckfeldt, P. E. Johnson, and J. D. Sprague, *Political Disagreement: The Survival of Diverse Opinions within Communication Networks* (New York: Cambridge University Press, 2004).

15. J. H. Fowler, "Turnout in a Small World," in *The Social Logic of Politics: Personal Networks as Contexts for Political Behavior*, ed. A. Zuckerman (Philadelphia: Temple University Press, 2005): 269–87.

16. R. Putnam, *Bowling Alone* (New York: Simon and Schuster, 2001).

17. D. W. Nickerson, "Is Voting Contagious? Evidence from Two Field Experiemtns," *American Political Science Review* 102 (2008): 49–57.

18. A. de Tocqueville, *Democracy in America*, trans. and ed. H. C. Mansfield and D. Winthrop (Chicago: University of Chicago Press, 2000).

19. B. C. Burden, "Voter Turnout and the National Election Studies," *Political Analysis* 8 (2000): 389–98.

20. K. T. Poole and H. Rosenthal, *Congress: A Political-Economic History of Roll Call Voting* (New York: Oxford University Press, 1997).

21. *Congressional Record* (Senate), September 11, 2006, S9297.

22. J. H. Fowler, "Legislative Cosponsorship Networks in the U.S. House and Senate," *Social Networks* 28 (2006): 454–65; J. H. Fowler, "Connecting the Congress: A Study of Cosponsorship Networks," *Political Analysis* 14 (2006): 456–87.

23. "Brazen Conspiracy," *Washington Post,* November 29, 2005.

24. Y. Zhang and others, "Community Structure in Congressional Networks," *Physica A* 387 (2008): 1705–12.

25. M. McGrory, "McCain, Gramm a Strange Pairing," *Omaha World Herald,* November 18, 1995.

26. J. Zengerle, "Clubbed," *New Republic,* May 7, 2001.

27. R. L. Hall, "Measuring Legislative Influence," *Legislative Studies Quarterly* 17 (1992): 205–31; B. Sinclair, *The Transformation of the U.S. Senate* (Baltimore, MD: Johns Hopkins University, 1989); S. Smith, *Call to Order* (Washington, DC: Brookings Institution, 1989); B. Weingast, "Fighting Fire with Fire: Amending Activity and Institutional Change in the Postreform Congress," in *The Post-Reform Congress*, ed. R. Davidson (New York: St. Martin's Press, 1991).

28. D. P. Carpenter, K. M. Esterling, and D. M. J. Lazer, "Friends, Brokers, and Transitivity: Who Informs Whom in Washington Politics?" *Journal of Politics* 66 (2004): 224–46; D. P. Carpenter, K. M. Esterling, and D. M. J. Lazer, "The Strength of Weak Ties in Lobbying Networks—Evidence from Health-Care Politics in the United States," *Journal of Theoretical Politics* 10 (1998): 417–44.

29. A. Hoffman, *Steal This Book* (New York: Grove Press, 1971).

30. M. T. Heaney and F. Rojas, "Partisans, Nonpartisans, and the Antiwar Movement in the United States," *American Politics Research* 35 (2007): 431–64.

31. Smith and Raine, "The Internet and the 2008 Election."

32. See, for example: L. A. Henao, "Columbians Tell FARC: 'Enough's enough'—In a March Organized on Facebook, Hundreds of Thousands Protested Against the Leftist Rebel Group Monday," *Christian Science Monitor,* February 6, 2008.

33. L. A. Adamic and N. Glance, "The Political Blogosphere and the 2004 U.S. Election: Divided They Blog," *Proceedings of the 3rd International Workshop on Link Discovery* (New York: Association for Computing Machinery, 2005), 36–43.

34. J. Kelly and B. Etling, "Mapping Iran's Online Public: Politics and Culture in the Persian Blogosphere," *Berkman Center Research Publication* 2008-01 (2008): 1–36.

35. Kelly and Etling, "Mapping," 6.

CHAPTER 7: IT'S IN OUR NATURE

1. "Survivor Recaps," http://www.cbs.com/primetime/survivor/recaps/?season=2 (accessed March 5, 2009).

2. B. Holldobler and E. O. Wilson, *The Superorganism: The Beauty, Elegance, and Strangeness of Insect Societies* (New York: W. W. Norton, 2009).

3. I. McEwan, *Enduring Love* (New York: Anchor Books, 1998).

4. R. Axelrod, *The Evolution Corporation* (New York: Basic Books, 1984).

5. C. Hauert and others, "Volunteering as Red Queen Mechanism for Cooperation in Public Goods Games," *Science* 296 (2002): 1129–32.

6. R. Boyd and P. J. Richardson, "Punishment Allows the Evolution of Cooperation (or Anything Else) in Sizable Groups," *Ethology and Sociobiology* 13 (1992): 171–95.

7. J. H. Fowler, "Altruistic Punishment and the Origin of Cooperation," *Proceedings of the National Academy of Sciences* 102 (2005): 7047–49.

8. C. Hauert and others, "Via Freedom to Coercion: The Emergence of Costly Punishment," *Science* 316 (2007): 1905–7.

9. J. S. Mill, *Essays on Some Unsettled Questions of Political Economy* (London: Longmans, Green, Reader, and Dyer, 1874): V.46.

10. W. Güth, R. Schmittberger, and B. Schwarze, "An Experimental Analysis of Ultimatum Bargaining," *Journal of Economic Behavior and Organization* 3 (1982): 367–88.

11. J. H. Fowler, "Altruism and Turnout," *Journal of Politics* 68 (2006): 674–83; J. H. Fowler and C. D. Kam, "Beyond the Self: Altruism, Social Identity, and Political Participation," *Journal of Politics* 69 (2007): 813–27; C. D. Kam, S. Cranmer, and J. H. Fowler, "When It's Not All About Me: Altruism, Participation, and Political Context" (unpublished paper); C. T. Dawes, P. J. Loewen, and J. H. Fowler, "Social Preferences and Political Participation" (unpublished paper).

12. "Cash Found in House's Walls Becomes Nightmare," Associated Press, November 8, 2008.

13. R. Frank, T. Gilovich, and D. Regan, "Does Studying Economics Inhibit Cooperation?" *Journal of Economic Perspectives* 7 (1993): 159–71.

14. J. Henrich, "Does Culture Matter in Economic Behavior? Ultimatum Game Bargaining Among the Machiguenga," *American Economic Review* 90 (2000): 973–79.

15. H. Xian and others, "Self-Reported Zygosity and the Equal-Environments Assumption for Psychiatric Disorders in the Vietnam Era Twin Registry," *Behavior Genetics* 30 (2000): 303–10; K. S. Kendler and others, "A Test of the Equal-Environment Assumption in Twin Studies of Psychiatric Illness," *Behavior Genetics* 23 (1993): 21–27; S. Scarr and L. Carter-Saltzman, "Twin Method: Defense of a Critical Assumption," *Behavior Genetics* 9 (1979): 527–42.

16. D. Cesarini and others, "Heritability of Cooperative Behavior in the Trust Game," *Proceedings of the National Academy of Sciences* 105 (2008): 3721–26.

17. J. H. Fowler, C. T. Dawes, and N. A. Christakis, "Model of Genetic Variation in Human Social Networks," *PNAS: Proceedings of the National Academy of Sciences* 106 (2009): 1720–24.

18. Worlds Collide Theory, http://www.urbandictionary.com/define.php?term=Worlds+Collide+Theory (accessed March 4, 2009).

19. D. I. Boomsma and others, "Genetic and Environmental Contributions to Loneliness in Adults: The Netherlands Twin Register Study," *Behavior Genetics* 35 (2005): 745–52.

20. Ibid.

21. M. M. Lim and others, "Enhanced Partner Preference in a Promiscuous Species by Manipulating the Expression of a Single Gene," *Nature* 429 (2004): 754–57.

22. A. Knafo and others, "Individual Differences in Allocation of Funds in the Dictator Game Associated with Length of the Arginine Vasopressin 1a Receptor Rs3 Promoter Region and Correlation Between Rs3 Length and Hippocampal mRNA," *Genes, Brain and Behavior* 7 (2008): 266–75.

23. J. C. Flack and others, "Policing Stabilizes Construction of Social Niches in Primates," *Nature* 439 (2006): 426–29.

24. K. Faust and J. Skvoretz, "Comparing Networks Across Space and Time, Size, and Species," *Sociological Methodology* 32 (2002): 267–99.

25. J. H. Fowler and D. Schreiber, "Biology, Politics, and the Emerging Science of Human Nature," *Science* 322 (2008): 912–14.

26. M. A. Changizi, Q. Zhang, and S. Shimojo, "Bare Skin, Blood and the Evolution of Primate Colour Vision," *Biology Letters* 2 (2006): 217–21.

27. E. Herrmann and others, "Humans Have Evolved Specialized Skills of Social Cognition: The Cultural Intelligence Hypothesis," *Science* 317 (2007): 1360–66.

28. C. Mamali, "Participative Pictorial Representations of Self-Other Relationships: Social-Autograph Method," paper presented at the General Meeting of the European Association of Experimental Social Pyschology, Croatia, June 1–14, 2008.

29. N. Epley and others, "Creating Social Connection Through Inferential Reproduction: Loneliness and Perceived Agency in Gadgets, Gods, and Greyhounds," *Psychological Science* 19 (2008): 114–20.

30. A. B. Newberg and others, "The Measurement of Regional Cerebral Blood Flow During the Complex Cognitive Task of Meditation: A Preliminary SPECT Study," *Psychiatry Research: Neuroimaging* 106 (2001): 113–22; A. B. Newberg and others, "Cerebral Blood Flow During Meditative Prayer: Preliminary Findings and Methodological Issues," *Perceptual and Motor Skills* 97 (2003): 625–30.

31. R. Dunbar, "Coevolution of Neocortex Size, Group Size, and Language in Humans," *Behavioral and Brain Sciences* 16 (1993): 681–735.

CHAPTER 8: HYPERCONNECTED

1. E. T. Lofgren and N. H. Fefferman, "The Untapped Potential of Virtual Game Worlds to Shed Light on Real World Epidemics," *Lancet Infectious Diseases* 7 (2007): 625–29.

2. S. Milgram, "Behavioral Study of Obedience," *Journal of Abnormal and Social Psychology* 67 (1963): 371–78; S. Milgram, *Obedience to Authority: An Experimental View* (New York: Harper Collins, 1974).

3. T. Blass, "The Milgram Paradigm After 35 years: Some Things We Now Know about Obedience to Authority," *Journal of Applied Social Psychology* 29 (1999): 955–78.

4. M. Slater and others, "A Virtual Reprise of the Stanley Milgram Obedience Experiments," *PLoS ONE* 1, no. 1 (2006): e39. doi:10.1371/journal.pone.0000039.

5. See, for example: A. Case, C. Paxson, and M. Islam, "Making Sense of the Labor Market Height Premium: Evidence from the British Household Panel Survey," NBER Working Paper 14007, May 2008; D. Hamermesh and J. Biddle, "Beauty and the Labor Market," *American Economic Review* 84 (1994): 1174–94; B. Harper, "Beauty, Stature and the Labour Market: A British Cohort Study," *Oxford Bulletin of Economics and Statistics* 62 (2008): 771–800; E. Loh, "The Economic Effects of Physical Appearance," *Social Science Quarterly* 74 (1993): 420–37.

6. N. Yee and J. Bailenson, "The Proteus Effect: The Effect of Transformed Self-Representation on Behavior," *Human Communication Research* 33 (2007): 271–90.

7. Ibid.

8. N. Yee, J. Bailenson, and N. Ducheneaut, "The Proteus Effect: Implications of Transformed Digital Self-Representation on Online and Offline Behavior," *Human Communication Research* 36 (2009): 285–312.

9. P. W. Eastwick and W. L. Garnder, "Is It a Game? Evidence for Social Influence in the Virtual World," *Social Influence* 1 (2008): 1–15.

10. N. Yee and others, "The Unbearable Likeness of Being Digital: The Persistence of Nonverbal Social Norms in Online Virtual Environments," *CyberPyschology and Behavior* 10 (2007): 115–21.

11. A. Cliff and P. Haggett, "Time, Travel, and Infection," *British Medical Bulletin* 69 (2004): 87–99.

12. Ibid.

13. D. J. Bradley, "The Scope of Travel Medicine" in *Travel Medicine: Proceedings of the First Conference on International Travel Medicine*, ed. R. Steffen (Berlin: Springer-Verlag, 1989): 1–9.

14. M. C. Gonzalez, C. A. Hidalgo, and A. L. Barabási, "Understanding Individual Human Mobility Patterns," *Nature* 453 (2008): 779–82.

15. T. Standage, *The Victorian Internet* (New York: Walker and Company, 1998).

16. I. de Sola Pool, *Forecasting the Telephone: A Retrospective Technology Assessment of the Telephone* (Norwood, NJ: Ablex Publishing, 1983): 86.

17. Ibid., 49.

18. C. S. Fischer, *America Calling: A Social History of the Telephone to 1940* (Berkeley, CA: University of California Press, 1992): 26.

19. C. H. Cooley, quoted in R. McKenzie, "The Neighborhood," reprinted in

Rodrick D. McKenzie on Human Ecology, ed. A. Hawley (Chicago: University of Chicago Press, 1921 [1968]): 51–93.

20. Fischer, *America Calling;* M. Mayer, "The Telephone and the Uses of Time," in *The Social Impact of the Telephone*, ed. I. de Sola Pool (Cambridge, MA: MIT Press, 1977), 225–45; N. S. Baron, *Always On: Language in an Online and Mobile World* (New York: Oxford University Press, 2008).

21. H. N. Casson, "The Social Value of the Telephone," *The Independent* 71 (1911): 899.

22. K. Hampton, "Netville: Community On and Offline in a Wired Suburb," in *The Cybercities Reader,* ed. S. Graham (London: Routledge, 2004): 260.

23. Hampton, "Netville," 260.

24. D. M. Boyd and N. B. Ellison, "Social Network Sites: Definition, History, and Scholarship," *Journal of Computer-Mediated Communication* 13 (2007): 210–30.

25. Ibid.

26. "Eliot Students Petition for Tape; Kirklanders Fast for Facebook," *Harvard Crimson,* December 1, 1984.

27. S. C. Faludi, "Help Wanted: Brass Tacks," *Harvard Crimson,* September 28, 1979.

28. "Facebook Statistics," http://www.facebook.com/press/info.php?statistics (accessed March 7, 2009)

29. K. Lewis and others, "Tastes, Ties, and Time: A New (Cultural, Multiplex, and Longitudinal) Social Network Dataset Using Facebook.com," *Social Networks* 30 (2008): 330–42.

30. *The Colbert Report,* July 31, 2006.

31. J. H. Fowler, "The Colbert Bump in Campaign Donations: More Truthful than Truthy," *PS: Political Science & Politics* 41 (2008): 533–39.

32. A. Ebersbach and others, *Wiki: Web Collaboration* (New York: Springer-Verlag, 2008).

33. J. Giles, "Internet Encyclopaedias Go Head to Head," *Nature* 438 (2005): 900–1.

34. "When Sam Met Allison," *Children's News,* September 2008.

35. "ACOR Acorlists," http://www.acor.org/about/about.html (accessed March 7, 2009).

36. J. D. Klausner and others, "Tracing a Syphilis Outbreak Through Cyberspace," *Journal of the American Medical Association* 284 (2000): 447–49.

37. A. Lenhart, L. Rainie, and O. Lewis, *Teenage Life Online: The Rise of the Instant-Message Generation and the Internet's Impact on Friendship and Family Relationships* (Washington, DC: Pew Internet and American Life Project, 2001).

38. A. Lenhart, M. Madden, and P. Hitlin, *Teens and Technology* (Washington, DC: Pew Internet and American Life Project, 2005).

39. J. L. Whitlock, J. L. Powers, and J. Eckenrode, "The Virtual Cutting Edge: The Internet and Adolescent Self-Injury," *Developmental Psychology* 42, no. 3 (2006): 1–11.

40. Ibid.

41. Ibid., 7.

42. http://gangstalkingworld.com (accessed November 6, 2008).

43. S. Kershaw, "Sharing Their Demons on the Web," *New York Times,* November 12, 2008; see also V. Bell, A. Munoz-Solomando, and V. Reddy, "'Mind Control' Experiences on the Internet: Implications for Psychiatric Diagnosis or Delusions," *Psychopathology* 39 (2006): 87–91.

44. "Woman Arrested for Killing Virtual Reality Husband," *CNN*, October 23, 2008.
45. "Virtual World Affair Ends with Real-Life Divorce," *Western Morning News*, November 14, 2008.

CHAPTER 9: THE WHOLE IS GREAT

1. Genesis, 11: 6 (King James Version).
2. T. Hobbes, *The Leviathan*, ed. M. Oakshott (Oxford: Oxford University Press, 1962): 100.
3. B. Holldobler and E. O. Wilson, *The Superorganism: The Beauty, Elegance, and Strangeness of Insect Societies* (New York: W. W. Norton, 2009).
4. M. Nowak, "Five Rules for the Evolution of Cooperation," *Science* 314 (2006): 1560–63.
5. I. D. Couzin and others, "Effective Leadership and Decision-Making in Animal Groups on the Move," *Nature* 433 (2005): 513–16; I. D. Couzin and others, "Collective Memory and Spatial Sorting in Animal Groups," *Journal of Theoretical Biology* 218 (2002): 1–11.
6. D. P. Bebber and others, "Biological Solutions to Transport Network Design," *Proceedings of the Royal Society B* 274 (2007): 2307–15; "Transport Efficiency and Resilience in Mycelial Networks," remarks by Mark Fricker at the Meeting of the German Physical Society, Dresden, March 27, 2009.
7. T. Nakagaki, H. Yamada, and A. Toth, "Maze-solving by an Amoeboid Organism," *Nature* 407 (2000): 470.
8. G. Palla, A. L. Barabási, and T. Vicsek, "Quantifying Social Group Evolution," *Nature* 446 (2007): 664–67.
9. S. Crabtree and B. Pelham, "Religion Provides Emotional Boost to World's Poor," March 6, 2009, http://www.gallup.com/poll/116449/Religion-Provides-Emotional-Boost-World-Poor.aspx.
10. E. L. Glaeser, B. Sacerdote, J. A. Scheinkman, "Crime and Social Interactions," *Quarterly Journal of Economics* 11 (1996): 507–48.
11. A. J. Reiss, "Understanding Changes in Crime Rates," in *Indicators of Crime and Criminal Justice: Quantitative Studies*, ed. S. Feinberg and A. J. Reiss (Washington, DC: Bureau of Justice Statistics, 1980).
12. F. Gino, S. Ayal, and D. Ariely, "Contagion and Differentiation in Unethical Behavior: The Effect of One Bad Apple on the Barrel," *Psychological Science* 20 (2009): 393–98.
13. Independent Sector, "Giving and Volunteering in the United States—2001," www.independentsector.org.
14. K. G. Carman, "Social Influences and the Private Provision of Public Goods: Evidence from Charitable Contributions in the Workplace," Stanford Institute for Economic Policy Research Discussion Paper 02-13, January 2003.
15. J. K. Goeree and others, "The 1/d Law of Giving," http://www.hss.caltech.edu/~lyariv/Papers/Westridge.pdf (accessed March 4, 2009).
16. S. Leider and others, "Directed Altruism and Enforced Reciprocity in Social Networks: How Much Is a Friend Worth?" (May 2007), NBER Working Paper No. W13135, http://ssrn.com/abstract=989946.

Illustration Credits

Plate 5: Adapted from M. Kearns, S. Suri, and N. Montfort, "An Experimental Study of the Coloring Problem on Human Subject Networks," *Science* 313 (2006): 824–27.

Plate 6: Reprinted with permission from L. A. Adamic and N. Glance, "The Political Blogosphere and the 2004 U.S. Election: Divided They Blog," *Proceedings of the 3rd International Workshop on Link Discovery* (New York: Association for Computing Machinery, 2005): 37.

Plate 7: Reprinted from J. Kelly and B. Etling, "Mapping Iran's Online Public," Berkman Center Research Publication 2008–01 (2008): 1–36.

Index

About the Authors

NICHOLAS A. CHRISTAKIS, MD, PhD, is both a physician and a social scientist. Medicine and social science are like distant cousins who meet from time to time, see that they have much in common, and then get into an argument. Christakis has been trying to referee these arguments for years, sometimes with surprising success. He is a professor of medical sociology in the Department of Health Care Policy at Harvard Medical School, professor of sociology in the Department of Sociology in the Harvard Faculty of Arts and Sciences, and professor of medicine and an attending physician in the Department of Medicine at Harvard Medical School. He was elected to the Institute of Medicine of the National Academy of Sciences in 2006. After finishing his clinical training in internal medicine, he combined his research with clinical practice as a hospice physician, looking for ways to improve end-of-life care. Since 1999, he has been investigating how social factors and social interactions affect health and longevity. Christakis is best known for his studies on how social networks form and operate. When he is not in the lab, he teaches students in many parts of Harvard University and in Harvard-affiliated hospitals, and is regularly voted a "favorite professor" by Harvard undergraduates because of his engaging lecture style and open office hours. He was named to *Time* magazine's list of the 100 most influential people in the world in 2009.

JAMES H. FOWLER, PhD, is a new kind of political scientist. In an effort to connect with the natural sciences, he pushes the boundaries of his field to identify social and biological forces that underlie human nature. Fowler's work as a Peace Corps volunteer in cholera-stricken villages in Ecuador motivated him to ask the question, why are some people so much better at facing group challenges than others? He studied politics at Harvard University where he earned his PhD, and he has since devoted his life to unifying the study of political outcomes with the study of other natural processes. He is currently an associate professor at the University of California, San Diego, in the Department of Political Science and the Center for Wireless and Population Health Systems. In addition to his work on social networks, Fowler is well known for his research on the evolution of cooperation, behavioral economics, political participation, and genopolitics (the study of the genetic basis of political behavior). While at Harvard he won several teaching awards, but now his students know him best as the professor who found the first scientific evidence for the "Colbert bump," a phenomenon in which political candidates tend to receive a boost in support after appearing on Stephen Colbert's comedy talk show.